TABLE OF CONTENTS

Fundamentals of Emergency Management

Independent Study 230.a
January 14, 2010

 FEMA

Table of Contents

Course Overview

Unit 1: Course Introduction
Introduction
How to Complete This Course
Unit 1 Objectives
Course Objectives
Case Study: Tornado in Barneveld, Wisconsin
Your Place in the Emergency Management System
Case Study: Hazardous Chemical Release
Activity: Where Do I Fit?

Unit 2: Overview of the Principles of Emergency Management and the Integrated Emergency Management System
Introduction and Unit Overview
FEMA Mission and Purpose
Response Authorities History
Principles of Emergency Management
Recent Changes to Emergency Planning Requirements
Why an Integrated Emergency Management System?
Emergency Management Concepts and Terms
Partners in the Coordination Network
Activity: Partners in the Coordination Network
Emergency Management in Local Government
Activity: Where Is Emergency Management in My Community?

Unit 3: Incident Management Actions
Introduction and Unit Overview
Introduction to the Spectrum of Incident Management Actions
Prevention
Preparedness
Response
Activity: Response Operations
Recovery
Mitigation

Unit 4: Roles of Key Participants
Introduction and Unit Overview
The Role of the Local Emergency Program Manager
State Emergency Management Role
How the Private Sector and Voluntary Organizations Assist Emergency Managers
Federal Emergency Management Role
The National Response Framework
Activity: Emergency Management Partners
Emergency Management Functional Groups
Case Study: Emergency Management Coordination

TABLE OF CONTENTS

Unit 5: The Plan as a Program Centerpiece
Introduction and Unit Overview
What Is an EOP and What Does It Do?
Activity: Where Do I Fit Into the EOP?
Case Study: An EOP in Action
Importance of the Hazard Analysis to the Planning Process
What Is In a Hazard Analysis?
Job Aid 5.1: Hazard Analysis Worksheet
Using a Hazard Analysis to Determine Risk
EOP Structure and Contents

Unit 6: Planning and Coordination
Introduction and Unit Overview
Linking Hazard Analysis to Capability Assessment
The EOP and the Incident Command System
The EOP and the EOC
Activity: The EOP, ICS, and EOC
Case Study: Multiple-Agency Coordination
Augmenting Local Resources
Maintaining an Effective EOP
Interfacing with Other Plans

Unit 7: Functions of an Emergency Management Program
Introduction and Unit Overview
Introduction to Emergency Management Functions
Basis in Local Law
Emergency Management Core Functions
Emergency Management Program Functions
Case Study: Train Derailment Review
Activity: Emergency Management Functions in Action
Activity: Comparing Functions

Unit 8: Applying Emergency Management Principles
Introduction and Unit Overview
Applying the Integrated Emergency Management System
Activity: Interdependence Within the Emergency Management Team
Activity: Problem Solving In Crisis-Prone County

Unit 9: Course Summary
Introduction and Unit Overview
Integrated Emergency Management System
The Spectrum of Incident Management Actions
The Plan as Program Centerpiece
Planning and Coordination
Functions of an Emergency Management Program
Emergency Management Program Partners
Applying Emergency Management Principles
Next Steps

Table of Contents

Final Exam

Appendix A: Job Aid
Appendix B: Acronym List
Appendix C: Emergency Supply Kit

Course Overview

COURSE OVERVIEW

About This Course

All communities are vulnerable to a variety of hazards. Emergency management provides a structure for anticipating and dealing with emergency incidents.

Emergency management involves participants at all governmental levels and in the private sector. Activities are geared according to phases before, during, and after emergency events. The effectiveness of emergency management rests on a network of relationships among partners in the system.

This course is one in the Federal Emergency Management Agency (FEMA) Professional Development Series. The goal of this course is to introduce you to the fundamentals of emergency management as an integrated system, surveying how the resources and capabilities of all functions at all levels can be networked together in all phases for all hazards.

Unit 1: Course Introduction

Unit 1: Course Introduction

Introduction

When an emergency or disaster strikes, you work as part of a complex emergency management network that calls upon many functions, resources, and capabilities. Your ability to function effectively relates to your understanding of how the emergency management system works and how your agency fits into the network. This course will present the fundamental aspects of emergency management and provide opportunities for you to apply what you learn.

Fundamentals of Emergency Management contains nine units. Each unit is described below.

- **Unit 1: Course Introduction,** offers an overview of the course content.

- **Unit 2: Overview of the Integrated Emergency Management System,** presents an overview of the integrated emergency management system.

- **Unit 3: The Spectrum of Incident Management Actions,** describes the phases of emergency management activities.

- **Unit 4: Roles of Key Participants,** examines the role of the local emergency program manager and relationships with State and Federal emergency managers.

- **Unit 5: The Plan as Program Centerpiece** focuses on community-specific risks and describes the hazard analysis process, and links hazard analysis to the EOP.

- **Unit 6: Planning and Coordination,** addresses resource requirements, how to supplement resources, the ICS-EOC interface, and the connection between planning and emergency management.

Unit 1: Course Introduction

How to Complete This Course (Continued)

- **Unit 7: Functions of an Emergency Management Program,** presents the core functions of an emergency management program.

- **Unit 8: Applying Emergency Management Principles** provides practice in applying emergency management principles in a problem-solving activity.

- **Unit 9: Course Summary and Final Exam,** summarizes the course content. At the conclusion of this unit, you will have an opportunity to evaluate the course and your success in meeting your personal learning goals. The final exam is also included in this unit.

Activities

This course will involve you actively as a learner by including activities that highlight basic concepts. It will also provide you with guidance on actions required in specific situations through the use of case studies. These activities emphasize different learning points, so be sure to complete all of them. Compare your answers to the answers provided following each activity. If your answers are correct, continue on with the material. If your answers are incorrect, go back and review the material before continuing.

Knowledge Checks

To help you know when to proceed to the next unit, Units 2 through 8 are followed by a Knowledge Check that asks you to answer questions that pertain to the unit content. The answers are given at the end of each knowledge check. When you finish each Knowledge Check, check your answers, and review the parts of the text that you do not understand. Do not proceed to the next unit until you are sure that you have mastered the current unit.

Appendices

In addition to the nine units, this course also includes three appendices. Appendix A includes a job aid, Appendix B includes an acronym list, and Appendix C includes the Emergency Supply Kit.

Unit 1: Course Introduction

Unit 1 Objectives

After completing this unit, you should be able to:

- Relate the topics to your job and community.
- Determine a strategy for completing the course successfully.

Course Objectives

This course is designed to introduce you to the fundamentals of emergency management. At the conclusion of this course, you should be able to:

- Organize emergency management functions, organizations, and activities using concepts and terms explained in the course.

- Explain the all-hazard emergency management process that integrates the resources of local, tribal, State, and Federal governments and voluntary and business assets.

- Explain the local, tribal, State, Federal, and individual and family roles in emergency management.

- Explain what individuals and families can do to protect themselves in emergencies.

- Describe the elements of an emergency management program.

- Discuss the role of individuals and organizations, as well as their relationships with one another, in emergency management.

- Explain the importance of networking to emergency management.

- Explain the social, political, and economic implications of a disaster.

- Describe alternate models for organizing emergency management programs.

UNIT 1: COURSE INTRODUCTION

Case Study: Tornado in Barneveld, Wisconsin

The case study on the following pages illustrates the need for emergency management. Read the case study and answer the questions that follow.

On June 8, 1984, at 12:50 a.m., a devastating tornado struck the small village of Barneveld, Wisconsin. Although a tornado watch was in effect, no warning was issued because the tornado originated near the town. The town, which had approximately 580 residents, was literally flattened by winds in excess of 200 miles an hour. Casualties add up to about 11% of the population: 9 lives were lost and 57 persons were treated for injuries. The storm destroyed 120 homes, 11 businesses, the village elementary school, 5 churches, and all of the municipal buildings, including a new fire station and the equipment in it. The village was left without electricity, telephone service, or water. Damage was estimated at over $20 million.

The local power company was in radio contact with the sheriff's office within 5 minutes and was moving trucks into the area within 40 minutes, encountering such hazardous conditions as exposed fuel oil and LP tanks. The telephone company set up an emergency bank of phones. Both companies needed several days to complete repairs. A command post was established to coordinate emergency operations. Local officials immediately began to clear debris from the stricken area. Police, fire, and emergency medical personnel concentrated their efforts on search and rescue operations for those who were trapped in collapsed structures. The village was evacuated to another town where congregate care was set up by the Red Cross, which also assisted in preliminary damage assessment.

The town received State assistance immediately. The State patrol directed traffic and assisted in securing portions of the affected area, and the National Guard assisted in security and law enforcement, as well as emergency operations. The Department of Natural Resources assisted in security, traffic control, and recovery operations. The State Department of Health and Social Services supported the county social service offices, which were quickly overwhelmed with requests for assistance. The State response was coordinated through the Emergency Operations Center, which was also dealing with other tornado damage.

The State requested Federal assistance on June 9, and it was granted. The disaster assistance center was located 20 miles from the town to serve survivors in other locations as well. Because few residents had cars in working order, transportation to the center was difficult. Many residents were angered to find that emergency loans required several months to process. Having no way to earn a living, many left the village.

The after-action plan noted that the county had no plan for debris removal, and that combustibles and non-combustibles should have been separated. There was no plan for a systematic turn-off of gas or for identification of hazardous materials and toxic substances. There was no plan designating who would be in charge of cleanup, although the highway commissioner eventually took this role. The best site for disposal had not been pre-designated. With 20-20 hindsight, officials realized that each county's emergency program manager should identify landfills in advance and mark out procedures for getting burning permits.

UNIT 1: COURSE INTRODUCTION

Case Study: Tornado in Barneveld, Wisconsin (Continued)

The town also lacked a plan to coordinate volunteer agencies. While there were many volunteers, no one was clearly in charge.

While our case study has focused on short-term effects, such a disaster can shatter a local economy and change the lives of residents for years. The emotional damage of living through such a disaster is less obvious than the physical devastation, but no less real. Providing emotional support to residents and helping them reconstruct their lives, including the economic base for their community, is a critical part of the recovery phase of any such emergency.

---

1. What effects can an emergency have on a small community—in this case, Barneveld, Wisconsin?

    - Financial

2. What kinds of emergency services are needed during and after an emergency?

3. What effects do emergency planning activities have on the response to a disaster, such as the tornado and recovery from it?

Your Place in the Emergency Management System

Normally, you work in a setting where day-to-day responsibilities are clear and lines of communication are well established through experience.

However, you also are part of a complex network of people and organizations responsible for dealing with emergencies in your local community. It is important that when the need arises, you know where you fit into that network and how to work within it.

UNIT 1: COURSE INTRODUCTION

Case Study: Hazardous Chemical Release

The case study that follows will help you think about where you fit into your community's emergency management network. Read the following description of a hazardous chemical release.

Think about what your role might be in such an incident. Your community may be exposed to a similar risk from hazardous chemical spills because of nearby rail lines, highways, or chemical plants or you may have responded to a similar incident.

When you have finished reading the case study, answer the questions that follow.

A freight train derailed in the upper Midwest in January 2002, in a county with a population of 60,000. Fifteen of the cars on the train contained anhydrous ammonia. (Anhydrous means "without water." Anhydrous ammonia seeks water from any source—even the human body. The compound will, therefore, seek the moisture in the eyes, nose, mouth, and lungs, causing caustic burns as it dissolves into body tissue. Inhaling large amounts of anhydrous ammonia will cause swelling of the throat and suffocation. Anhydrous ammonia is transported as a liquid under pressure.)

When the train derailed, eight of the fifteen cars ruptured, causing an explosion of the pressurized chemical. The force of the explosion sent one piece of a car slicing into a house a mile away. The blast caused the release of 240,000 to 290,000 gallons of anhydrous ammonia gas—the largest release in the world to date.

The incident occurred at 3:00 AM, when warning systems that rely on radio or television transmission fail to alert most people. Any evacuation attempt would have exposed residents to greater hazard. As a result, they were initially advised to shelter in their homes. Eventually 21 homes were evacuated. One resident died while attempting to leave the area. Approximately one third of a nearby city was also affected, but residents were not able to evacuate. Those affected were advised to shelter in place.

There were also some delays in activating responders, who could not enter the accident vicinity without proper gear. Fire-fighting gear does not offer adequate protection. One responder was trapped after he drove into a ditch trying to leave the scene because his vehicle windshield was coated with frozen gas in the toxic cloud. The responder was rescued some time later.

Residents were told to turn off their furnaces to avoid drawing outside air into their homes. Public heath was a major concern. Approximately 400 patients were processed through hospitals.

Unit 1: Course Introduction

Case Study: Hazardous Chemical Release (Continued)

Media attention was intense. Individuals and families needed public information on treating exposure symptoms, cleaning homes, and dealing with exposed pets and livestock. Many horses, being especially sensitive to airborne contaminants, died.

The cause for the derailment is still unknown, but was probably a faulty rail or wheel on the train. Possible sabotage has not been ruled out.

UNIT 1: COURSE INTRODUCTION

Activity: Where Do I Fit?

Many agencies are involved in such an incident. Emergency management, fire and police departments from different jurisdictions, voluntary agencies, emergency medical and health officials, and environmental agencies are among those to respond or deal with the aftermath of a hazardous materials release, such as that described in the scenario.

Think about what your department or agency would have done if the derailment and subsequent chemical release had happened in your community, and answer the questions below.

1. What role would your agency, department, or organization play during and after this incident?

    - Containment

2. What types of duties would you be likely to perform?

    Support Haz-Mat Team
    De Con Citizens +
    Evacuate

3. Name three points of contact that would be important to completing your responsibilities successfully during and after such an incident in your community.

    - Chemtrec
    - Haz Mat Team Leader
    - IC

Unit 2: Overview of the Principles of Emergency Management and the Integrated Emergency Management System

UNIT 2: OVERVIEW of the Principles of Emergency Management and the Integrated Emergency Management System

Introduction and Unit Overview

This unit will provide an overview of an integrated emergency management system, and where you fit within the system.

After completing this unit, you should be able to:

- State FEMA's Mission and Purpose
- Describe the history of National response authorities
- Describe the 8 Emergency Management Principles
- Describe the integrated emergency management system and what the system should do.
- Define emergency management concepts and terms.
- Identify the players in the emergency management network.
- Describe the roles of the key players in the emergency management system.
- Identify the location of the emergency management function within their local government.

UNIT 2: OVERVIEW of the Principles of Emergency Management and the Integrated Emergency Management System

FEMA Mission and Purpose

In 1979, the Federal Emergency Management Agency (FEMA) was established by an executive order, which merged many of the separate disaster-related responsibilities into a single agency. FEMA became part of the Department of Homeland Security (DHS) in 2003.

FEMA's mission is to reduce the loss of life and property and protect communities nationwide from all hazards, including natural disasters, acts of terrorism, and other human caused disasters. FEMA leads and supports the nation in a risk-based, comprehensive emergency management system of preparedness, protection, response, recovery and mitigation.

One main goal of FEMA is to provide disaster assistance to individuals and communities. FEMA does not assume total responsibility for disaster assistance but does assume the role of coordinating Federal, State, and local efforts when a Federal emergency or disaster is declared.

Response Authorities: A brief history

The role of the Federal Government in disaster response has evolved throughout the past 200 years. The Congressional Act of 1803 was the earliest effort to provide disaster relief on a federal level after a fire devastated a New Hampshire town. From that point forward, assorted legislation provided disaster support. Between 1803 and 1950, the Federal Government intervened in approximately 100 incidents (earthquakes, fires, floods, and tornadoes). The Federal Government became more proactive based on the following authorities:

- The Defense Production Act of 1950 was the first comprehensive legislation pertaining to Federal disaster relief.

- In 1952, President Truman issued Executive Order 10427, which emphasized that **Federal disaster assistance was intended to supplement, not supplant, the resources of State, local, and private-sector organizations.**

UNIT 2: OVERVIEW of the Principles of Emergency Management and the Integrated Emergency Management System

The Stafford Act

Today, the centerpiece legislation for providing Federal aid in disaster relief is the Robert T. Stafford Disaster Relief and Emergency Assistance Act (Public Law 100-707). The Stafford Act:

- Provides a system of emergency preparedness for the protection of life and property from hazards.
- Vests responsibility for emergency preparedness jointly in the Federal Government, State governments, and their political subdivisions.
- Gives FEMA responsibility for coordinating Federal Government response.

Under the Stafford Act, assistance is limited to:

- Natural catastrophe (including any hurricane, tornado, storm, high water, wind driven water, tidal wave, tsunami, earthquake, volcanic eruption, landslide, mudslide, snowstorm, or drought), or,
- Regardless of cause, any fire, flood, or explosion.

The Stafford Act is designed to supplement the efforts and available resources of States, tribes, local governments, and disaster relief organizations.

Under the Stafford Act, the President can designate an incident either as an "emergency" or a "major disaster." Both authorize the Federal Government to provide essential assistance to meet immediate threats to life and property, as well as additional disaster relief assistance.

The President may, in certain circumstances, declare an "emergency" unilaterally, but may only declare a "major disaster" at the request of a Governor who certifies the State and affected local governments are overwhelmed.

Unit 2: Overview of the Principles of Emergency Management and the Integrated Emergency Management System

Stafford Act: Emergency and Major Disaster Declarations

Types of Incidents

- **Emergency.** Emergencies involve any event for which the President determines there is a need to supplement State and local efforts in order to save lives, protect property and public health, and ensure safety. A variety of incidents may qualify as emergencies. The Federal assistance available for emergencies is more limited than that available for a major disaster.
  An emergency is defined as, "Any occasion or instance for which, in the determination of the President, Federal assistance is needed to supplement State and local efforts and capabilities to save lives and to protect property and public health and safety, or lessen or avert the threat of a catastrophe in any part of the United States."

  A Presidential declaration of an emergency provides assistance that:

  - Is beyond State and local capabilities.
  - Serves as supplementary emergency assistance.
  - Does not exceed $5 million of Federal assistance.

  The Governor of an affected State must request a Presidential Declaration for an emergency within 5 days of the incident.

- **Major disaster.** Major disasters may be caused by such natural events as floods, hurricanes, and earthquakes. Disasters may include fires, floods, or explosions that the President feels are of sufficient magnitude to warrant Federal assistance. Although the types of incidents that may qualify as a major disaster are limited, the Federal assistance available for major disasters is broader than that available for emergencies.

  A major disaster is defined as, "Any natural catastrophe ... or, regardless of cause, any fire, flood, or explosion, in any part of the United States, which in the determination of the President causes damage of sufficient severity and magnitude to warrant major disaster assistance under this chapter to supplement the efforts and available resources of States, local governments, and disaster relief organizations in alleviating the damage, loss, hardship, or suffering caused thereby."

  A Presidential disaster declaration provides assistance that:

  - Is beyond State and local capabilities.

- Supplements available resources of State and local governments, disaster relief organizations, and insurance.

The Governor of an affected State must request a Presidential declaration for a major disaster within 30 days of the incident. Additional information on the request process is presented later in this section.

Emergency Management Principles

In March of 2007, Emergency Management Institute's Higher Education Project convened a working group of emergency management practitioners and academics to consider principles of emergency management. This project was prompted by the realization that while numerous books, articles and papers referred to "principles of emergency management", nowhere in the literature was there an agreed upon definition of these principles.

The group agreed on eight principles that will be used to guide the development of a doctrine of emergency management.

Principles

The groups agreed Emergency Management must be:

1. **Comprehensive** - emergency managers consider and take into account all hazards, all phases, all stakeholders and all impacts relevant to disasters.

2. **Progressive** - emergency managers anticipate future disasters and take protective, preventive and preparatory measures to build disaster-resistant and disaster-resilient communities.

3. **Risk-Driven** - emergency managers use sound risk management principles (hazard identification, risk analysis, and impact analysis) in assigning priorities and resources.

4. **Integrated** - emergency managers ensure unity of effort among all levels of government and all elements of a community.

5. **Collaborative** - emergency managers create and sustain broad and sincere relationships among individuals and organizations to encourage trust, advocate a team atmosphere, build consensus, and

UNIT 2: OVERVIEW of the Principles of Emergency Management and the Integrated Emergency Management System

facilitate communication.

6. **Coordinated** - emergency managers synchronize the activities of all relevant stakeholders to achieve a common purpose.

7. **Flexible** - emergency managers use creative and innovative approaches in solving disaster challenges.

8. **Professional** - emergency managers value a science and knowledge-based approach based on education, training, experience, ethical practice, public stewardship and continuous improvement.

Recent Changes to Emergency Planning Requirements

The terrorist attacks of September 11, 2001, illustrated the need for all levels of government, the private sector, and nongovernmental agencies to prepare for, protect against, respond to, and recover from a wide spectrum of events that exceed the capabilities of any single entity. These events require a unified and coordinated national approach to planning and to domestic incident management. To address this need, the President signed a series of Homeland Security Presidential Directives (HSPDs) that were intended to develop a common approach to preparedness and response. The HSPDs include:

- HSPD-5, Management of Domestic Incidents, identifies steps for improved coordination in response to incidents. It requires the Department of Homeland Security (DHS) to coordinate with other Federal departments and agencies and State, local, and tribal governments to establish a National Response Plan (NRP) and a National Incident Management System (NIMS). NOTE: In January 2008, DHS issued the National Response Framework (NRF) that supersedes the NRP.

- HSPD-8, National Preparedness, describes the way Federal departments and agencies will prepare. It requires DHS to coordinate with other Federal departments and agencies—and with State, local, and tribal governments to develop a National Preparedness Goal.

The complimentary nature of NIMS and the NRF define what needs to be done to prevent, protect against, respond to, and recover from a major event. These efforts align Federal, State, local, and tribal entities; the private sector; and nongovernmental agencies to provide an effective and efficient national structure for preparedness, incident management, and emergency response.

Unit 2: Overview of the Principles of Emergency Management and the Integrated Emergency Management System

Recent Changes to Emergency Planning Requirements (Continued)

NIMS

Each day communities respond to numerous emergencies. Most often, these incidents are managed effectively at the local level.

However, there are some incidents that may require a collaborative approach that includes personnel from:

- Multiple jurisdictions,
- A combination of specialties or disciplines,
- Several levels of government,
- Nongovernmental organizations, and
- The private sector.

The National Incident Management System, or NIMS, provides the foundation needed to ensure that we can work together when our communities and the Nation need us the most.

NIMS integrates best practices into a comprehensive, standardized framework that is flexible enough to be applicable across the full spectrum of potential incidents, regardless of cause, size, location, or complexity.

Using NIMS allows us to work together to prepare for, prevent, respond to, recover from, and mitigate the effects of incidents.

NIMS provides a consistent framework for incident management at all jurisdictional levels, regardless of the cause, size, or complexity of the incident. Building on the Incident Command System (ICS), NIMS provides the Nation's first responders and authorities with the same foundation for incident management for terrorist attacks, natural disasters, and all other emergencies. NIMS requires that ICS be institutionalized and used to manage all domestic incidents.

At the policy level, institutionalizing ICS means that government officials:

- Adopt ICS through executive order, proclamation or legislation as the jurisdiction's official incident response system.
- Direct that incident managers and response organizations in their jurisdictions train, exercise, and use ICS in their response operations.

At the organizational/operational level, incident managers and emergency response organizations should:

- Integrate ICS into functional, system-wide emergency operations policies, plans, and procedures.
- Provide ICS training for responders, supervisors, and command-level officers.
- Conduct exercises for responders at all levels, including responders from all disciplines and jurisdictions.

Unit 2: Overview of the Principles of Emergency Management and the Integrated Emergency Management System

NIMS (continued)

NIMS integrates existing best practices into a consistent, nationwide approach to domestic incident management that is applicable at all jurisdictional levels and across functional disciplines.

Following is a synopsis of each major component of NIMS.

- **Preparedness.** Effective incident management and incident response activities begin with a host of preparedness activities conducted on an ongoing basis, in advance of any potential incident. Preparedness involves an integrated combination of planning, procedures and protocols, training and exercises, personnel qualification and certification, and equipment certification.

- **Communications and Information Management.** Emergency management and incident response activities rely on communications and information systems that provide a common operating picture to all command and coordination sites. NIMS describes the requirements necessary for a standardized framework for communications and emphasizes the need for a common operating picture. NIMS is based on the concepts of interoperability, reliability, scalability, portability, and the resiliency and redundancy of communications and information systems.

- **Resource Management.** Resources (such as personnel, equipment, and/or supplies) are needed to support critical incident objectives. The flow of resources must be fluid and adaptable to the requirements of the incident. NIMS defines standardized mechanisms and establishes the resource management process to: identify requirements for, order and acquire, mobilize, track and report, recover and demobilize, reimburse for, and inventory resources.

- **Command and Management.** The Command and Management component within NIMS is designed to enable effective and efficient incident management and coordination by providing flexible, standardized incident management structures. The structure is based on three key organizational constructs: the

    Incident Command System
    Multiagency Coordination Systems
    Public Information.

- **Ongoing Management and Maintenance.** DHS/FEMA manages the development and maintenance of NIMS. This includes developing NIMS programs and processes as well as keeping the NIMS document current.

UNIT 2: OVERVIEW of the Principles of Emergency Management and the Integrated Emergency Management System

National Response Framework (NRF)

The *National Response Framework (NRF)* is a guide to how the Nation conducts all-hazards response. It is built upon *scalable, flexible, and adaptable coordinating structures* to align key roles and responsibilities *across the Nation*. It describes specific authorities and best practices for managing incidents that range from the serious but purely local, to large-scale terrorist attacks or catastrophic natural disasters.

The NRF explains the common discipline and structures that have been exercised and matured at the local, tribal, State, and national levels over time. It describes key lessons learned from Hurricanes Katrina and Rita, focusing particularly on how the Federal Government is organized to support communities and States in catastrophic incidents.

The term "response" in the *Framework* includes immediate actions to save lives, protect property and the environment, and meet basic human needs. Response also includes the execution of emergency plans and actions to support short-term recovery. The *Framework* is always in effect, and elements can be implemented as needed on a flexible, scalable basis to improve response.

The *Framework* is written especially for government executives, private-sector and nongovernmental organization (NGO) leaders, and emergency management practitioners. First, it is addressed to senior elected and appointed leaders, such as Federal department or agency heads, State Governors, mayors, tribal leaders, and city or county officials – those who have a responsibility to provide for effective response. For the Nation to be prepared for all hazards, its leaders must have a baseline familiarity with the concepts and mechanics of the *Framework*.

- Codification of FEMA's Role – PKEMRA clarified FEMA's role in incidents by changing the Homeland Security Act. See pages 98 – 103 in small Stafford Act booklet. 6 USC 313 – 316, especially Section 503 of the Homeland Security Act, 6 USC 313(b) (2) (A).

The **Emergency Management Accreditation Program (EMAP)** is a standard-based voluntary assessment and accreditation process for government programs responsible for coordinating prevention, mitigation, preparedness, response, and recovery activities for natural and human-caused disasters. Accreditation is based on compliance with collaboratively developed national standards, the *Emergency Management Standard by EMAP.*

The *Emergency Management Standard by EMAP* is designed as a tool for continuous improvement as part of a voluntary accreditation process for local and state emergency management programs.

Unit 2: Overview of the Principles of Emergency Management and the Integrated Emergency Management System

EMAP Purpose

The goal of EMAP is to provide a meaningful, voluntary accreditation process for state, territorial, and local programs that have the responsibility of preparing for and responding to disasters. By offering consistent standards and a process through which emergency management programs can demonstrate compliance, EMAP will strengthen communities' capabilities in responding to all types of hazards, from tornadoes and earthquakes to school violence and bioterrorism. Accreditation is voluntary. Its intent is to encourage examination of strengths and weaknesses, pursuit of corrective measures, and communication and planning among different sectors of government and the community.

National emergency management and other stakeholder organizations began working on what is now EMAP in 1997 after a presentation on the need for emergency management standards and accreditation at a National Emergency Management Association (NEMA) conference. EMAP builds on standards and assessment work by various organizations, adding requirements for documentation and verification that neither standards nor self-assessment alone can provide. EMAP consists of:

Agreed-upon national standards (*Emergency Management Standard*) developed with input from emergency managers and state and local government officials include;

- Program Management
- Administration and Finance
- Laws and Authorities
- Hazard Identification, Risk Assessment and Consequence Analysis
- Hazard Mitigation
- Prevention and Security
- Planning
- Incident Management
- Resource Management and Logistics
- Mutual Aid
- Communications and Warning
- Operations and Procedures
- Facilities
- Training

UNIT 2: OVERVIEW of the Principles of Emergency Management and the Integrated Emergency Management System

- Exercises, Evaluations and Corrective Action
- Crisis Communications, Public Education and Information

The Legal Basis for Continuity of Operations

National Security Presidential Directive-51 (NSPD-51)/Homeland Security Presidential Directive-20 (HSPD-20), National Continuity Policy, specifies certain requirements for continuity plan development, including the requirement that all Federal executive branch departments and agencies develop an integrated, overlapping continuity capability.

NFPA 1600 Standard on Disaster/Emergency Management and Business Continuity Programs 2007 Edition

The NFPA Standards Council established the Disaster Management Committee in January 1991. The committee was given the responsibility for developing documents relating to preparedness for, response to, and recovery from disasters resulting from natural, human, or technological events.

The first document that the committee focused on was NFPA1600, *Recommended Practice for Disaster Management*. NFPA 1600 was presented to the NFPA membership at the 1995 Annual Meeting in Denver, CO. That effort produced the 1995 edition of NFPA 1600.

For the 2000 edition, the committee incorporated a "total program approach" for disaster/emergency management and business continuity programs in its revision of the document from a recommended practice to a standard. They provided a standardized basis for disaster/emergency management planning and business continuity programs in private and public sectors by providing common program elements, techniques, and processes. The committee provided expanded provisions for enhanced capabilities for disaster/emergency management and business continuity programs so that the impacts of a disaster would be mitigated, while protecting life and property. The chapters were expanded to include additional material relating to disaster/emergency management and business continuity programs. The annex material was also expanded to include additional explanatory material.

UNIT 2: OVERVIEW of the Principles of Emergency Management and the Integrated Emergency Management System

For the 2004 edition, the committee updated terminology and editorially reformatted the document to follow the 2003 *Manual of Style for NFPA Technical Committee Documents*; however, the basic features of the standard remained unchanged. In addition, the committee added a table in Annex A that created a crosswalk among FEMA CAR, NFPA 1600, and the Business Continuity Institute and Disaster Recovery Institute International professional practices. The committee added significant informational resources to Annexes B, C, D, and E. The document continues to be developed in cooperation and coordination with representatives from FEMA, NEMA, and IAEM. This coordinated effort was reflected in the expansion of the title of the standard for the 2000 edition to include both disaster and emergency management, as well as information on business continuity programs.

The 2007 edition incorporates changes to the 2004 edition, expanding the conceptual framework for disaster/emergency management and business continuity programs. Previous editions of the standard focused on the four aspects of mitigation, preparedness, response, and recovery. The 2007 edition identifies prevention as a distinct aspect of the program, in addition to the other four. Doing so brings the standard into alignment with related disciplines and practices of risk management, security, and loss prevention.

The proposed 2010 standards address:
- Leadership and Commitment
- Program Coordinator
- Program Committee
- Program Administration
- Performance Objectives
- Laws and Authorities
- Finance and Administration
- Records Management
- Planning and Design
- Risk Assessment
- Incident Prevention
- Mitigation
- Planning Process
- Common Plan Requirements
- Resource Management
- Mutual Aid/Assistance
- Communications and Warning
- Operational Procedures
- Emergency Response
- Business Continuity and Recovery
- Crisis Communications, Public Information, and Education
- Incident Management
- Emergency Operations Centers (EOCs)
- Training and Education
- Testing and Exercises
- Program Improvement
- Program Review

Unit 2: Overview of the Principles of Emergency Management and the Integrated Emergency Management System

- Corrective Action

UNIT 2: OVERVIEW of the Principles of Emergency Management and the Integrated Emergency Management System

What These Changes Mean to You

Depending on your jurisdiction, the changes to the emergency planning requirements may mean little—or a lot. Minimally, the changes mean that your jurisdiction must:

- Use ICS to manage all incidents, including recurring and/or planned special events.

- Integrate all response agencies into a single, seamless system, from the Incident Command Post, through Department Emergency Operations Centers (DEOCs) and local Emergency Operations Centers (EOCs), through the State EOC to the regional- and national-level entities.

- Develop and implement a public information system.

- Identify and type all resources according to established standards.

- Ensure that all personnel are trained properly for the job(s) they perform.

- Ensure communications interoperability and redundancy.

Consider each of these requirements in the context of the principles of emergency management as presented in this course.

UNIT 2: OVERVIEW of the Principles of Emergency Management and the Integrated Emergency Management System

Why an Integrated Emergency Management System?

When an emergency or disaster occurs:

- Personnel from different agencies, jurisdictions, and governmental levels need to work together.

- Quick decisions are required.

To facilitate rapid, efficient emergency operations, a system is required that enables all participants in the incident to work together. An **integrated emergency management system** is a conceptual framework to increase emergency management capability by networking. That increased capability would not be readily available, especially in a disaster, without establishing prior networking, coordination, linkages, interoperability, partnerships, and creative thinking about resource shortfalls. The system should address all hazards that threaten a community, be useful in all four phases of emergency management, seek resources from any and all sources that are appropriate, and knit together all partnerships and participants for a mutual goal.

Emergency Management Concepts and Terms

Many emergency management terms are used throughout this course. To avoid confusion, this course establishes a single definition for each term. These may differ from how you use the terms in your community.

For clear reference during the course, however, please use the definitions on the pages that follow.

Disaster:
An occurrence of a natural catastrophe, technological accident, or human-caused event that has resulted in severe property damage, deaths, and/or multiple injuries. As used in this course, a "large-scale disaster" is one that exceeds the response capability of the Local jurisdiction and requires State, and potentially Federal, involvement. As used in the Stafford Act, a "major disaster" is "any natural catastrophe [...] or, regardless of cause, any fire, flood, or explosion, in any part of the United States, which in the determination of the President causes damage of sufficient severity and magnitude to warrant major disaster assistance under [the] Act to supplement the efforts and available resources of States, local governments, and disaster relief organizations in alleviating the damage, loss, hardship, or suffering caused thereby." (Stafford Act, Sec. 102(2), 42 U.S.C. 5122(2)).

UNIT 2: OVERVIEW of the Principles of Emergency Management and the Integrated Emergency Management System

Emergency Management Concepts and Terms (Continued)

Emergency: Any incident, whether natural or manmade, that requires responsive action to protect life or property. Typically, emergencies can be handled at the local level.

However, under the Robert T. Stafford Disaster Relief and Emergency Assistance Act, an emergency "means any occasion or instance for which, in the determination of the President, Federal assistance is needed to supplement State and local efforts and capabilities to save lives and to protect property and public health and safety, or to lessen or avert the threat of a catastrophe in any part of the United States" (Stafford Act, Sec. 102(1), 42 U.S.C. 5122(1)).

Emergency Management: Organized *analysis, planning, decision-making, and assignment of available resources* to mitigate, prepare for, respond to, and recover from the effects of all hazards.

The goals of emergency management are to:

- Save lives.
- Prevent injuries.
- Protect property and the environment.

Hazard: Something that is potentially dangerous or harmful, often the root cause of an unwanted outcome. Natural hazards are caused by natural events that pose a threat to lives, property, and other assets. Technological hazards are caused by the tools, machines, and substances we use in everyday life. Intentional hazards, such as terrorism or riots, are deliberately caused by people attacking or damaging what is valuable in a society.

UNIT 2: OVERVIEW of the Principles of Emergency Management and the Integrated Emergency Management System

Partners in the Coordination Network

Effective response to and recovery from an emergency or disaster requires the active involvement of numerous partners.

Government Partners

Each level of government participates in and contributes to emergency management.

- Local government has direct responsibility for the safety of its people, knowledge of the situation and accompanying resource requirements, and proximity to both event and resources. Within local government are Emergency Support Services—the departments of local government that are capable of responding to emergencies 24 hours a day. They include law enforcement, fire/rescue, and public works. They may also be referred to as emergency response personnel or first responders.

- Tribal Government - **The United States has a trust relationship with Indian tribes** and recognizes their right to self-government. As such, tribal governments are responsible for coordinating resources to address actual or potential incidents. When local resources are not adequate, tribal leaders seek assistance from States or the Federal Government.

  For certain types of Federal assistance, tribal governments work with the State, but as sovereign entities, they can elect to deal directly with the Federal Government for other types of assistance. In order to obtain Federal assistance via the Stafford Act, **a State Governor must request a Presidential declaration on behalf of a tribe**.

- State government has legal authorities for emergency response and recovery and serves as the point of contact between local and Federal governments.

- Federal government has legal authorities; fiscal resources; research capabilities, technical information and services; and specialized personnel to assist local and State agencies in responding to and recovering from emergencies or disasters.

Organizations at all government levels can share their knowledge and resources with nongovernmental service providers. For example:

- At the local level, first-response agencies share information about injuries with local medical providers. Information about those who are left homeless from a disaster is shared with The American Red Cross and other community service organizations.

Unit 2: Overview of the Principles of Emergency Management and the Integrated Emergency Management System

- At the <u>State level</u>, the Governor's Authorized Representative (GAR) and others share information with State agencies (e.g., Department of Agriculture) and FEMA regional representatives to bring the necessary response and recovery resources to bear on the incident.

- At the tribal level, the tribal leader is responsible for the public safety and welfare of the people of that tribe. As authorized by tribal government, the tribal leader:

    • Communicates with the tribal community, and helps people, businesses, and organizations cope with the consequences of any type of incident.

    • Can request Federal assistance under the Stafford Act through the Governor of the State when it becomes clear that the tribe's capabilities will be insufficient or have been exceeded.

    • Can elect to deal directly with the Federal Government. Although a State Governor must request a Presidential declaration on behalf of a tribe under the Stafford Act, Federal departments or agencies can work directly with the tribe within existing authorities and resources.

- At the <u>Federal level</u>, when an incident occurs that exceeds or is anticipated to exceed local or State resources – or when an incident is managed by Federal departments or agencies acting under their own authorities – the Federal Government uses the National Response Framework to involve all necessary department and agency capabilities, organize the Federal response, and ensure coordination with response partners.

Unit 2: Overview of the Principles of Emergency Management and the Integrated Emergency Management System

Private Sector Partners

Government agencies are responsible for protecting life and property and promoting wellbeing. But the government does not—and cannot—work alone.

In all facets of emergencies and disasters, the government works with private-sector groups as partners in emergency management.

The term **private sector** includes *nongovernmental organizations (NGOs) that offer critical emergency services, such as The American Red Cross, as well as businesses that have resources to contribute.*

Together, government agencies and the private sector form a partnership. This partnership begins at the grassroots level, depending on the local and State resources that are in place, to provide the backbone for disaster management. Humanitarian and volunteer organizations also are essential to the team.

NGOs collaborate with first responders, governments at all levels, and other agencies and organizations providing relief services to sustain life, reduce physical and emotional distress, and promote recovery of disaster survivors.

The National Voluntary Organizations Active in Disasters (NVOAD) is a consortium of more than 30 recognized national organizations active in disaster relief. Such organizations provide capabilities to incident management and response efforts at all levels.

The private sector (both for-profit and nonprofit entities):

- Bears the greatest casualties and costs of emergencies.
- Provides voluntary expertise and support for emergency management.

The private sector makes its concerns known to the government, and holds the government accountable for actions taken or not taken. Regardless of government accountability, communities could not respond to or recover from emergencies or disasters without the assistance of and cooperation from the private sector.

Unit 2: Overview of the Principles of Emergency Management and the Integrated Emergency Management System

Individuals and families as Partners

Although not formally a part of emergency management, individuals and families play an important role in the overall emergency management process. Private individuals and families can contribute by:

- Reducing hazards in and around their homes. By taking simple actions, such as raising utilities above flood levels or taking in objects that could become projectiles in a high wind, individuals and families can reduce the amount of damage caused by an emergency or disaster event.

- Preparing a disaster supply kit. By assembling disaster supplies in advance of an event, individuals and families can take care of themselves until first-responders arrive. (See the recommended disaster supplies list in Appendix C to this course.)

- Monitoring emergency communications carefully. Throughout an emergency situation, critical information and direction will be released to the public via electronic and other media. By listening and following these directions carefully, individuals and families can reduce their risk of injury, keep emergency routes open to response personnel, and reduce demands on landline and cellular communication.

- Volunteering with an established organization. Organizations and agencies with a role in emergency response and recovery are always seeking hard-working, dedicated volunteers. By volunteering with an established voluntary agency, individuals and families can become part of the emergency management system and assure that their efforts are directed to where they are most needed.

- Taking training in emergency response. Taking training in emergency response, whether the training is basic first aid through The American Red Cross or a more complex course through a local community college, will enable individuals and families to take initial response actions required to take care of themselves and their families, thus freeing first-response personnel to respond to higher-priority incidents that affect the entire community.

- Community Emergency Response Team (CERT) training is one way for individuals and families to prepare for an emergency. CERT is designed to prepare people to help themselves, their families, and their neighbors in the event of a catastrophic disaster. Because emergency services personnel may not be able to help everyone immediately, people can make a difference by using the training obtained in the CERT course to save lives and protect property.

Individuals and families as Partners (Continued)

This training covers basic disaster survival and rescue skills that are important to have in a disaster when emergency services are not available. Some of the topics covered are:

- Disaster preparedness—anticipating the impact on an infrastructure, safety precautions during a disaster, and the role of CERTs in disaster response.
- Basic fire safety—identifying and reducing potential fire hazards, how to evaluate fires, and firefighting resources and techniques (e.g., use of portable fire extinguishers).
- Disaster medical operations—principles of triage, assessment of injuries, and treatment.
- Light search and rescue operations—priorities and resources; lifting, cribbing, and victim removal; and rescuer safety.
- Disaster psychology team organization—the psychological impact of a disaster on rescuers and survivors, and how to provide psychological "first aid."

Additional courses of interest are available through FEMA's Emergency Management Institute home study program (http://training.fema.gov/EMIWeb/). Your State may also offer training opportunities through its Emergency Management Agency.

UNIT 2: OVERVIEW of the Principles of Emergency Management and the Integrated Emergency Management System

Activity: Partners in the Coordination Network

The purpose of this activity is to ensure that you understand the functions of key participants in emergency management.

For each participant in Column A, choose a description from Column B.

Matching Participant to Description
1. __B__ Emergency support services   a. Acts as a liaison between local and Federal authorities
2. __E__ Private sector
b. Includes law enforcement, fire/rescue, and public works
3. __D__ Local government
4. __A__ State government   c. May offer fiscal resources, technical assistance, and specialized personnel
5. __C__ Federal government
d. Has proximity to the event and resources
e. Experiences the greatest casualties and costs of disasters

Fundamentals of Emergency Management IS 230.a

UNIT 2: OVERVIEW of the Principles of Emergency Management and the Integrated Emergency Management System

Answers

Activity: Emergency Management Participants (Continued)

1. b
2. e
3. d
4. a
5. c

Emergency Management in Local Government

Resources for an integrated emergency management system include both personnel and equipment.

Personnel resources in your area include:

- Elected and appointed officials and executives.
- Emergency program managers.
- Emergency operations staff.
- Police and fire departments.
- Other local service providers, such as the local council on aging and public works agency.
- Voluntary organizations such as The American Red Cross and The Salvation Army.

An integrated emergency management system links these personnel resources through:

- Planning.
- Direction.
- Coordination.
- Clearly defined roles and functions.

A successful emergency management program facilitates the development of a network of relationships among local officials and staff that understand their roles and are able to act when needed.

The organizational placement of emergency management affects the way that relationships are developed.

Emergency Management in Local Government (Continued)

Where is the emergency management function in your local government's organization chart? Some options include placing it within:

- A separate organization that reports directly to a governing or executive body.

- The fire/rescue department.

- Law enforcement, located in a police department or sheriff's office.

Separate Emergency Management Organization

An advantage of working within a separate organization is that the perception of bias is minimized. The emergency management function may become more visible and have increased access within local government.

A disadvantage of working within a separate organization is that the emergency management staff must work to build rapport and avoid becoming isolated.

Placement Within Fire/Rescue or Law Enforcement Departments

These agencies are among the traditional first responders to emergencies and disasters, so placing the emergency manager within a first-response agency is logical.

An advantage of working within a first-response agency is that being close to the day-to-day operations of law enforcement or fire personnel builds personal relationships that pay off in coordination when developing and maintaining an emergency management program.

A disadvantage of working within a first-response agency is that association with one or another of these basic services may hamper coalition-building efforts if others perceive the emergency management staff as owing allegiance to its own service.

UNIT 2: OVERVIEW of the Principles of Emergency Management and the Integrated Emergency Management System

Activity: Where Is Emergency Management in My Community?

This activity will provide you with an opportunity to explore the emergency management functions in your community. Please take some time to research your local emergency management functions. Then, answer the questions below.

1. The local emergency management function is located (organizationally):

   ☒ As part of the fire department.
   ☐ As part of law enforcement.
   ☐ As an independent agency.

2. Who is the local emergency manager?

   Fire Chief

3. To whom does the emergency manager report?

   Mayor

4. What are the advantages and disadvantages of this reporting relationship? Also, think about recent emergency responses. How do you think the emergency management's organization facilitated that response?

UNIT 2: OVERVIEW of the Principles of Emergency Management and the Integrated Emergency Management System

Knowledge Check

Carefully read each question and all of the possible answers before selecting the most appropriate response for each test item. Circle the letter corresponding to the answer you have chosen.

1. A(n) _____ is defined as a dangerous event or circumstance that has the potential to lead to an emergency or disaster:

   a. **Hazard**
   b. Chemical spill
   c. Emergency activation
   d. Drought
   e. Power outage

2. In emergency management, personnel are considered one type of resource.

   a. **True**
   b. False

3. One goal of emergency management is to:

   a. Predict and minimize damage resulting from earthquakes.
   b. Conduct exercises based on simulated incidents.
   c. Supplement State and local efforts and capabilities.
   d. **Identify hazards.**
   e. Prevent injuries resulting from hazards.

4. A local emergency manager and staff often serve as a function of which department in the local government?

   a. Finance department
   b. Public works department
   c. Planning commission
   d. **Fire/rescue service**
   e. Volunteer coordination office

5. An emergency management program will work well in practice if most emphasis and attention focus upon _____.

   a. A comprehensive written plan
   b. **Well-established, day-to-day relationships**
   c. Reliance on State assistance
   d. Mutual aid and assistance

Fundamentals of Emergency Management IS 230.a

UNIT 2: OVERVIEW of the Principles of Emergency Management and the Integrated Emergency Management System

Knowledge Check (Continued)

Answers to Questions

1. a
2. a
3. d
4. d
5. b

Unit 3: Incident Management Actions

Unit 3: Incident Management Actions

Introduction and Unit Overview

This unit examines incident management actions, including the five phases of emergency management activities. After completing this unit, you should be able to:

- Describe the five phases of emergency management activities.
- Identify measures that the communities, individuals and families can take in connection with each of the five phases.
- Describe the planning activities and documents that pertain at the local, tribal, State, and Federal levels.
- Identify the types of assistance that may be available from the Federal government.

Introduction to Incident Management Actions

An emergency management program examines potential emergencies and disasters based on the risks posed by likely hazards; develops and implements programs aimed toward reducing the impact of these events on the community, prepares for those risks that cannot be eliminated; and prescribes the actions required to deal with the consequences of actual events and to recover from those events.

- Pre-incident activities, such as information sharing, threat identification, planning, and readiness exercises.
- Incident activities that include lifesaving missions and critical infrastructure support protections.
- Post-incident activities that help people and communities recover and rebuild for a safer future.

Unit 3: Incident Management Actions

Introduction to Incident Management Actions (Continued)

Incident management actions illustrated below. Traditional definitions of the terms have been expanded to include pre-incident activities.

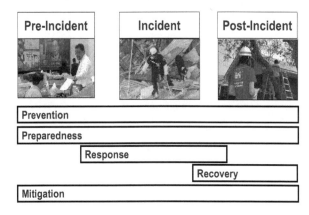

Each of the phases will be described in more detail in the next sections.

Unit 3: Incident Management Actions

Prevention

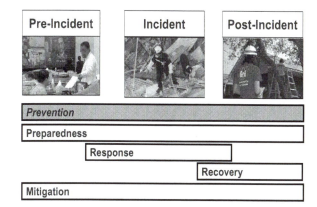

Prevention:

- Means actions taken to avoid an incident or to intervene to stop an incident from occurring.

- Involves actions taken to protect lives and property.

- Involves applying intelligence and other information to a range of activities that may include such countermeasures as:

    - Deterrence operations
    - Heightened inspections
    - Improved surveillance
    - Interconnections of health and disease prevention among people, domestic animals, and wildlife.

Unit 3: Incident Management Actions

Preparedness

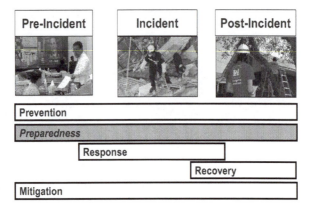

Because it is not possible to prevent or mitigate completely against every hazard that poses a risk, **preparedness** measures can help to reduce the impact of the remaining hazards by taking certain actions before an emergency event occurs. Preparedness includes *plans or other preparations made to save lives and facilitate response and recovery operations.*

Preparedness is defined as the range of deliberate, critical tasks and activities necessary to build, sustain, and improve the operational capability to prevent, protect against, respond to, and recover from domestic incidents. Preparedness is a continuous process involving efforts at all levels of government and between government and private-sector and nongovernmental organizations to identify threats, determine vulnerabilities, and identify required resources.

Preparedness measures involve all of the players in the integrated emergency management system—local, tribal, State, and Federal agencies and individuals and families—and, at the local level, may include activities, such as:

- Developing an Emergency Operations Plan (EOP) that addresses identified hazards, risks, and response measures.

- Recruiting, assigning, and training staff who can assist in key areas of response operations.

- Identifying resources and supplies that may be required in an emergency.

- Designating facilities for emergency use.

UNIT 3: INCIDENT MANAGEMENT ACTIONS

The EOP

Generally, the EOP describes how the community (or State) will do business in an emergency. The EOP:

- Assigns responsibility to organizations and individuals for carrying out specific actions that exceed the capability or responsibility of any single agency.

- Establishes lines of authority and organizational relationships, and shows how all actions will be coordinated.

- Describes how people and property will be protected in emergencies and disasters.

- Identifies personnel, equipment, facilities, supplies, and other resources that can be made available—within the jurisdiction or by agreement with other jurisdictions—for use during response and recovery operations.

- Identifies steps to address mitigation concerns during response and recovery operations.

Local government is responsible for attending to the public's emergency needs. Therefore, the local EOP focuses on measures that are essential for protecting the public, including:

- **Warning and communications:** How the local government will warn the public of an existing or impending emergency and communicate internally before, during, and after an event occurs.

- **Emergency public information:** How government will communicate with the public before, during, and after an emergency occurs. Emergency public information is especially critical in light of the recent terrorism threat. Decisions about what to tell the public and when are critical to gaining a reasoned response from the public, providing confidence that the government is doing all it can to protect the public and control the situation, and—perhaps most importantly—making the public into a response asset will be crucial.

- **Mass care:** Where and for how long the public's emergency needs, such as shelter and food distribution, will be accomplished. What facilities will be available, what supplies will be stocked, and how the supplies will be distributed are all covered under mass care in the EOP.

The EOP (Continued)

- **Health and medical care:** How survivors will be cared for, where, and by who are addressed in the health and medical portion of the EOP. Special issues, such as decontamination, must also be addressed for hazardous materials and terrorist events.

- **Evacuation:** What routes will be used if evacuation becomes necessary, special transportation or routing requirements (e.g., evacuating the disabled or making evacuation routes one way to accommodate increased traffic flow), and other issues dealing with emergency egress are all part of the evacuation portion of the EOP.

States also have EOPs. State EOPs serve three main purposes:

1. To facilitate a State first response to certain emergencies.

2. To assist local jurisdictions during emergencies in which local response capabilities are overwhelmed.

3. To serve as a liaison with the Federal government in cases where Federal assistance is necessary and authorized.

The State EOP establishes the framework within which local EOPs are created and through which the Federal government becomes involved in response and recovery operations. As such, the State government acts as the coordinating entity to ensure that all levels of government are able to respond to safeguard the well-being of its people.

More information on the EOPs as they relate to preparedness is included in Unit 5 of this course.

Unit 3: Incident Management Actions

The EOP (Continued)

Recruiting, Assigning, and Training Staff

During an emergency or disaster response, it may be necessary to assign personnel to jobs other than those that they normally perform. Some personnel may already be employed within the community, but others may be recruited specifically for service in emergencies. Regardless of employment status, these personnel must be recruited, assigned, and trained for their jobs *before* an emergency event occurs. Whenever possible, these persons should be included in exercises that enable them to practice the job under simulated emergency conditions so that, when an actual emergency occurs, they are ready to perform in their new capacities with little or no time lost in learning the job.

Identifying Resources and Supplies

Identifying the resources and supplies that will be available for an emergency response is a crucial part of preparedness. Virtually all jurisdictions take an inventory of their personnel and equipment resources to determine what they have and compare it with what they may need in an emergency. Those gaps between on-hand resources and probable requirements can be filled in a number of ways. Among the most common are:

Identifying Resources and Supplies (Continued)

- **Mutual aid and assistance agreements** with neighboring jurisdictions. Mutual aid agreements are formal, written agreements between jurisdictions that provide the conditions under which resource sharing can take place during an emergency. These are most common among fire departments and law enforcement agencies but may be developed to cover other resources and equipment (e.g., construction equipment) as well.

- **Standby contracts** with suppliers of critical equipment and supplies. Standby contracts typically are made for equipment, such as dump trucks or other construction equipment, but are also used for supplies, such as plastic sheeting. Under a typical standby contract, the supplier agrees to provide an established quantity of an item at the unit cost in effect on the day *before* the emergency occurs. Standby contracts are a good way for local governments to meet their resource supply requirements without incurring the costs of stockpiling and without paying the rapidly increasing prices that often follow an emergency.

- **Resource typing** is designed to enhance emergency readiness and response at all levels of government through a comprehensive and integrated system that allows jurisdictions to augment their response resources during an incident. Specifically, it allows emergency management personnel to identify, locate, request, order, and track outside resources quickly and effectively and facilitate the response of these resources to the requesting jurisdiction.

In some large emergencies, State and Federal resources may be available. For example, the National Guard may be activated following an extremely heavy snow, in the case of wildfire, or following a terrorist incident. Federal resources, including Disaster Medical Assistance Teams (DMATs) and Disaster Mortuary Teams (DMORTs) may be activated following a mass-casualty incident. All requests for State and Federal resources must be processed through the State.

Unit 3: Incident Management Actions

Designating Facilities for Emergency Use

To ensure an effective and efficient response, certain facilities are designated as part of the emergency planning process. Typically, these facilities include:

- The <u>Emergency Operations Center (EOC)</u>, which is the central location from which all off-scene activities are coordinated. Senior elected and appointed officials are located at the EOC, as well as personnel supporting critical functions, such as operations, planning, logistics, and finance and administration. The key function of EOC personnel is to ensure that those who are located at the scene have the resources (i.e., personnel, tools, and equipment) they need for the response. In large emergencies and disasters, the EOC also acts as a liaison between local responders and the State. (Note that States operate EOCs as well and can activate them as necessary to support local operations. State EOC personnel report to the Governor and act as a liaison between local and Federal personnel.)

- <u>Shelters</u>, which are used to house survivors. Shelters should be designated *before* an event occurs, and the public should be aware of shelter locations and transportation routes from their neighborhoods or workplaces to the shelters. In most areas, The American Red Cross operates shelters and coordinates with the local volunteer program manager to ensure that sheltering needs are met.

- <u>Distribution centers</u>, from which food and emergency supplies are made available to the public. In most areas, The American Red Cross, together with other local voluntary agencies, coordinate distribution centers.

- <u>Storage areas</u> for specific types of equipment. Warehouses, supply yards, and other facilities that will be used as providers of the equipment necessary for a response should be designated as part of the planning process.

Other facilities may also be designated in advance, based on the jurisdiction's resources and the areas of the community that are likely to be affected. On-scene facilities, such as the Incident Command Post (ICP) and staging areas, typically are *not* designated in advance because of the requirement for close proximity to the incident site.

Unit 3: Incident Management Actions

Text Telephone (TTY) Alert: Lee County Division of Public Safety, Fort Myers, Florida

TTY Alert is an emergency warning system for deaf and hard-of-hearing residents in northwest Florida. It is the first system of its kind in the United States. When an emergency occurs, the Lee County EOP sends out an alert to the TTY machines with information about the emergency and information about what to do to every registered TTY user in the county. If necessary, the system can target a specific area. TTY Alert also allows TTY users to access the system to obtain headline news, weather bulletins, and family disaster preparedness information.

TTY Alert has been well received by the hearing-impaired community and has been recognized by the National Institute on Disabilities Rehabilitation Research.

Local Emergency Management/Industry Partnership: St. Charles Parish, Louisiana

The local emergency management/industry partnership program offers a telephone hotline system to coordinate response to disasters and emergencies. The program was established by the St. Charles Parish EOC in cooperation with 26 petrochemical companies. The system serves as a 24-hour warning system, an emergency information exchange, and a link between the companies and the parish Department of Emergency Preparedness for support during emergencies.

This system has been recognized by the Chemical Manufacturers Association as a model of government–industry cooperation.

> Source: *Partnerships in Preparedness: A Compendium of Exemplary Practices in Emergency Management*, Federal Emergency Management Agency, December 1995

Preparedness covers a range of activities and can be taken at all levels of government. Some examples that have been cited as being particularly effective for individuals and families are shown on the next page.

Individual and family Preparedness

Individuals and families can and should prepare for emergencies. There are several simple steps that you can take to prepare yourself for an emergency. Personal preparedness activities can not only keep you and your family safe but can help you become a response asset rather than a response burden.

- **Complete your own hazard analysis.** If you have lived in the community for any period of time, you are probably aware of the hazards that are high risk for your area. If you are new to the area, talk to some long-time residents to determine what events have occurred historically in your area. Don't forget the "small" emergencies, such as fire or an extended electrical outage.

- **Develop your own emergency plan.** Play the "what if" game with each of the hazards you selected. *What* would you do *if* _____ occurs? Then ask yourself what supplies you would need to take the action(s) you identify, and gather the supplies together.

- **Practice your plan.** Even simple tasks can become difficult during an emergency. **Practice your plan** *before* **an emergency** occurs until you are thoroughly familiar with the procedures you need to follow if the event occurs.

Response

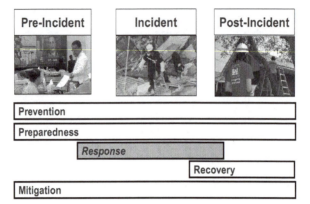

Response begins when an emergency event is imminent or immediately after an event occurs. Response encompasses the activities that address the short-term, direct effects of an incident. Response also includes the execution of EOPs and of incident mitigation activities designed to limit the loss of life, personal injury, property damage, and unfavorable outcomes.

As indicated by the situation, response activities include:

- Applying intelligence and other information to lessen the effects or consequences of an incident.

- Increasing security operations.

- Continuing investigations into the nature and source of the threat.

- Ongoing public health and agricultural surveillance and testing processes, immunizations, isolation, or quarantine.

- Specific law enforcement operations aimed at preempting, interdicting, or disrupting illegal activity, and apprehending actual perpetrators and bringing them to justice.

- Restoring critical infrastructure (e.g., utilities).

- Ensuring continuity of critical services (e.g., law enforcement, public works).

Response (Continued)

In other words, response involves putting preparedness plans into action.

One of the first response tasks is to conduct a situation assessment. Local government is responsible for emergency response and for continued assessment of its ability to protect its people and the property within the community. To fulfill this responsibility, responders and local government officials must conduct an immediate **rapid assessment** of the local situation.

Rapid assessment includes all immediate response activities that are directly linked to determining initial lifesaving and life-sustaining needs and to identifying imminent hazards. The ability of local governments to perform a rapid assessment within the first few hours after an event is crucial to providing an adequate response for life-threatening situations and imminent hazards. Coordinated and timely assessments enable local government to:

- Prioritize response activities.
- Allocate scarce resources.
- Request additional assistance from mutual aid partners, as well as the State, quickly and accurately.

Obtaining accurate information quickly through rapid assessment is key to initiating response activities and needs to be collected in an organized fashion. Critical information, also called **essential elements of information (EEI),** includes information about:

- Lifesaving needs, such as evacuation and search and rescue.
- The status of critical infrastructure, such as transportation, utilities, communication systems, and fuel and water supplies.
- The status of critical facilities, such as police and fire stations, medical providers, water and sewage treatment facilities, and media outlets.
- The risk of damage to the community (e.g., dams and levees, facilities producing or storing hazardous materials) from imminent hazards.
- The number of individuals who have been displaced because of the event and the estimated extent of damage to their dwellings.

Response (Continued)

Essential elements of information also include information about the potential for **cascading events.** Cascading events are events that occur as a direct or indirect result of an initial event. For example, if a flash flood disrupts electricity to an area and, as a result of the electrical failure, a serious traffic accident involving a hazardous materials spill occurs, the traffic accident is a cascading event. If, as a result of the hazardous materials spill, a neighborhood must be evacuated and a local stream is contaminated, these are also cascading events. Taken together, the effect of cascading events can be crippling to a community.

Good planning, training, and exercising before an event occurs can help reduce cascading events and their effects. Maintaining the discipline to follow the plan during response operations also reduces the effects of cascading events.

Individuals and families and Response Operations

What can individuals and families do to facilitate an emergency response? Surprisingly, there is much that people can do, and many of the actions that will help the response most are relatively simple.

- Follow your own emergency plan. Assuming that you developed a plan and practiced what you would do during the preparedness phase, this is the time to implement it. Follow your plan unless something related to the event makes it unworkable or unsafe.

- Pay attention to *and follow* emergency directions provided by local officials. Listen to emergency broadcasts on the local media and follow the directions provided in the broadcasts. Emergency announcements are prepared by those who are most familiar with what is actually happening at the incident site and will provide you with the information you need to remain safe during the emergency.

- Don't make unnecessary phone calls, either by cellular phone or land line. Keep critical lines of communication open for emergency use.

Very importantly, if you think you want to help during an emergency, don't just show up at the scene to help. Volunteer with an established voluntary agency *now*. Volunteering before an emergency occurs will enable you to receive the training you need so that, when an emergency occurs and your services are needed, you know where you need to go and what you will do. Volunteering before an emergency also helps the agency and local authorities identify their resources and plan their needs.

UNIT 3: INCIDENT MANAGEMENT ACTIONS

Activity: Response Operations

This activity provides you with the opportunity to reflect on past response operations in your community. To complete this activity, read and respond to the questions below.

1. Think about a recent emergency event that occurred in your community. What types of damage occurred as a result of the event?

2. Were you involved in the response? If yes, what was your job?

3. What do you think worked well with the response?

4. If the situation occurred again, what would you do differently (or what would you want local officials to do differently)?

5. List ways in which you think that preparedness activities contributed to the response.

Recovery

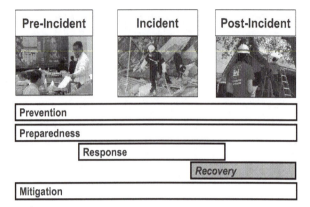

The goal of recovery is to return the community's systems and activities to normal. Recovery begins right after the emergency. Some recovery activities may be concurrent with response efforts.

Recovery is the development, coordination, and execution of service- and site-restoration plans for impacted communities and the reconstitution of government operations and services through individual, private-sector, nongovernmental, and public assistance programs that:

- Identify needs and define resources.
- Provide housing and promote restoration.
- Address long-term care and treatment of affected persons.
- Implement additional measures for community restoration.
- Incorporate mitigation measures and techniques, as feasible.
- Evaluate the incident to identify lessons learned.
- Develop initiatives to mitigate the effects of future incidents.

Long-term recovery includes restoring economic activity and rebuilding community facilities and housing. Long-term recovery (stabilizing all systems) can sometimes take years.

Recovery (Continued)

Although recovery is primarily a responsibility of local government, if the emergency or disaster received a Presidential Declaration, a number of assistance programs may be available under the Stafford Act. There are two major categories of Federal aid: Public Assistance and Individual Assistance.

Public Assistance is for repair of infrastructure, public facilities and debris removal, and may include.

- Repair or replacement of non-Federal roads, public buildings, and bridges.
- Implementation of Mitigation measures.

Individual Assistance is for damage to residences and businesses or for personal property losses, and may include:

- Grants to individuals and families for temporary housing, repairs, replacement of possessions, and medical and funeral expenses.
- The Small Business Administration (SBA) loans to individuals and businesses.
- Crisis counseling for survivors and responders; legal services; and disaster unemployment benefits.

Recovery from disaster is unique to each community depending on the amount and kind of damage caused by the disaster and the resources that the community has ready or can get. In the short term, recovery is an extension of the response phase in which basic services and functions are restored. In the long term, recovery is a restoration of both the personal lives of individuals and the livelihood of the community.

After the short-term recovery when roads have been opened, debris removed, supplies and shelters secured, communication channels, water and power, life safety and other basic services restored, the community and its leadership must rebuild.

Recovery (Continued)

Long term recovery may take several months or even extend into years because it is a complex process of revitalizing not just homes but also businesses, public infrastructure, and the community's economy and quality of life.

There are many long term leadership and planning considerations. Applying for assistance programs available from the Federal government, as mentioned previously, is important to consider for obtaining financial and other resources in the case of a Presidential Disaster Declaration. Other considerations include:

- Keeping people informed and preventing unrealistic expectations.
- Mitigation measures to ensure against future disaster damage.
- Donations
- Partnerships with business and industry for resources.
- Competing interests of groups involved in the planning process.
- Environmental issues.
- Public health measures to take against the risks of diseases, contamination, and other cascading effects from a disaster.
- The unmet needs of survivors.
- Rebuilding bridges, roads, public works, and other parts of the infrastructure.

Mitigation

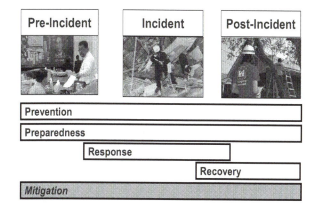

Mitigation refers to activities that are designed to:

- Reduce or eliminate risks to persons or property, or
- Lessen the actual or potential effects or consequences of an incident.

Mitigation measures:

- May be implemented prior to, during, or after an incident.
- Are often developed in accordance with lessons learned from prior incidents.
- Can include efforts to educate governments, businesses, and the public on measures that they can take to reduce loss and injury.

Mitigation is accomplished in conjunction with a **hazard analysis** (which will be covered in Unit 4). A hazard analysis helps to identify:

- What events can occur in and around the community?
- The likelihood that an event will occur.
- The consequences of the event in terms of casualties, destruction, disruption to critical services, and costs of recovery.

To be successful, mitigation measures must be developed into an overall **mitigation strategy** that considers ways to reduce hazard losses together with the overall risk from specific hazards and other community goals.

Developing a Mitigation Strategy

A sound mitigation strategy is one that is based on several factors:

- Prevention measures are intended to prevent existing risks from becoming worse based on new development or other changes within the community (e.g., road construction, zoning or building code changes). Prevention measures can be very effective in areas that have not been developed or are in an early phase of development. By implementing prevention measures, such as open space preservation and storm water management, future development can be directed in such a way as to minimize the risk from known hazards while maintaining other community goals and the overall quality of life in the community.

- Property protection measures are used to modify buildings or their surroundings to reduce the risk of damage from a known hazard. Property protection measures directly protect people and property at risk and may be simple and relatively low cost (e.g., raising utilities or strapping water heaters) or they may be more elaborate and expensive (e.g., acquiring land and using that land for recreational purposes or building earthquake-resistant structures in earthquake zones).

- Natural resource protection measures are used to reduce the consequences of a known hazard and to improve the overall quality of the environment. Natural resource protection measures can range from erosion and sediment control to wetlands protection to controlling runoff from farmland sediment into downstream waterways.

- Emergency services measures protect people before and after an event occurs and may include:

  - Warning.
  - Response.
  - Protective measures for critical facilities.
  - Maintenance of health and safety.

To be effective, emergency protective measures should be built into the emergency planning process, exercised, and revised to incorporate lessons learned from both exercises and actual emergencies.

Developing a Mitigation Strategy (Continued)

- **Structural projects** directly protect people and property that are at risk from a known hazard. Structural projects involve the construction of man-made structures (e.g., dikes, levees, elevated roadways) to control the damage from a known hazard. These projects can be very expensive, and over the long-term, may actually disrupt the environment in such a way as to increase the overall risk from other hazards. Additionally, some structural mitigation measures may provide the public with a false sense of security, especially in the case of an extreme event, such as the Midwest floods, during which many levees were breached by the flood waters.

- **Public information** serves to inform and remind people about the hazards they face and measures they should take to avoid damage or injury. Public information measures may include:
  - Outreach projects.
  - Real estate disclosure requirements.
  - Technical assistance.
  - Education programs.

The mitigation strategy developed must consider the hazards faced, the potential for damage from those hazards, and the overall needs of the community. Mitigation measures must be consistent with the strategy but can be effective only if considered as part of the larger emergency management cycle.

Mitigation measures can be developed and implemented at the local or State level. Two examples of mitigation measures that have been cited for their effectiveness are included below.

Unit 3: Incident Management Actions

Hazard Minimization Program: State of Massachusetts

The Hazard Minimization Program was instituted in November 1991, as a way to reduce repetitive losses from disasters. The program funds mitigation measures, such as basement window replacement and installation of interior flood walls as a way of reducing repetitive losses from flooding. To be eligible for the minimization program, individuals or families must have suffered a loss that can be minimized through a one-time mitigation measure.

Following a major storm in December 1992, the State conducted a survey to measure the program's success. Of the 71 homeowners who responded to the survey, 49 (69 percent) indicated that they had homes exposed to floodwaters from the storm, but only 3 (less than 1 percent) were affected by the floodwaters.

Of the three homes affected by the floodwaters, only one case related to a failed minimization project. During a follow-up survey, most participants stated that they would have been affected by the storm had minimization measures not been undertaken. These survey results indicated that the program could have a major impact on reducing future storm losses, both in terms of human suffering and in taxpayer dollars saved.

Hazard Mitigation Program: Borough of Avalon, Cape May County, New Jersey

The Borough of Avalon, Cape May County, New Jersey developed a mitigation strategy designed to minimize the impact of storm damage through the implementation of structural and nonstructural mitigation projects. Projects completed under the program included:

- Developing flood-level maps and installing flood-level indicators at specific points in the borough. These maps were then mass-mailed, together with a letter of explanation, to all borough residents.
- Preparing and distributing a quarterly newsletter to inform residents of emergency management proposals, such as evacuation routes, dredging and beach-fill projects, and shelter locations.
- Preparing a hazard mitigation plan for the borough, including goals and objectives, proposed measures, programs, and actions to avoid vulnerability to hazards and overall beach protection strategies.
- Conducting educational seminars in the borough on measures, procedures, and problems related to severe weather emergencies; distributing informational material; and creating an instructional videotape.
- Adopting land use and development ordinances and funding appropriations for property development restrictions; maintaining beaches, including installing sand fencing, planting dune grass, and implementing beach re-nourishment projects.

UNIT 3: INCIDENT MANAGEMENT ACTIONS

Hazard Mitigation Program: Borough of Avalon, Cape May County, New Jersey (Continued)

- Elevating the municipal building, police headquarters, and public works garage above the base flood elevation.
- Installing a borough wide public address warning system that includes television access through the local cable television company.

This strategy was awarded First Place for municipalities by the 1993 National Coordinating Council on Emergency Management.

Source: *Partnerships in Preparedness: A Compendium of Exemplary Practices in Emergency Management,* Federal Emergency Management Agency, December 1995

Unit 3: Incident Management Actions

Knowledge Check

Carefully read each question and all of the possible answers before selecting the most appropriate response for each test item. Circle the letter corresponding to the answer you have chosen.

1. Each phase of the emergency management cycle ends before the next one begins.

    a. True
    b. False

2. One example of mitigation is:

    a. Preparing a home disaster kit.
    b. Ordering evacuation.
    c. Learning cardiopulmonary resuscitation.
    d. Passing an ordinance controlling development in a floodplain.

3. The five phases of emergency management are useful for:

    a. Limiting activities to certain times.
    b. Keeping staff within boundaries.
    c. Prioritizing actions and resources.
    d. Providing categories to organize similar activities.

4. Response imposes the greatest time pressures on emergency management.

    a. True
    b. False

5. Federal assistance is available for which of the following purposes?

    a. Mitigation measures
    b. Medical and funeral expenses
    c. Temporary housing
    d. All of the above

Fundamentals of Emergency Management IS 230.a   Page 3.24

Knowledge Check (Continued)

6. Match each of the following activities with the emergency management phase when the activity would occur from the phase from the activity list.

   Activity
   a. Conducting a training exercise
   b. Floodplain mapping
   c. Opening shelters
   d. Rebuilding roads

   Phase
   _B_ Mitigation
   _D_ Recovery
   _A_ Preparedness
   _C_ Response

UNIT 3: INCIDENT MANAGEMENT ACTIONS

Knowledge Check (Continued)

Answers to Knowledge Check

1. b
2. d
3. d
4. a
5. d
6. b, d, a, c.

Unit 4: Roles of Key Participants

Unit 4: Roles of Key Participants

Introduction and Unit Overview

In the previous unit, you learned about emergency management core functions and program functions. After completing this unit you should be able to:

- Describe the role of the local Emergency Program Manager.
- Describe the tribal role in emergency management
- Discuss the State's emergency management role.
- Describe how private sector and voluntary organizations assist emergency managers.
- Discuss the Federal role in emergencies through the National Response Framework (NRF).
- List the emergency management functional groups.

The Role of the Local Emergency Program Manager

The local Emergency Program Manager has the day-to-day responsibility of overseeing emergency management programs and activities. And most emergencies are handled at the local level without State or Federal assistance. This role entails coordinating all aspects of a jurisdiction's mitigation, preparedness, response, and recovery capabilities. The Emergency Program Manager:

- Coordinates resources from all sectors before, during, and after an emergency.
- Manages activities relating to mitigation, preparedness, response, and recovery.
- Ensures that all players of the process:
    - Are aware of potential threats to the community.
    - Participate in mitigation and prevention activities.
    - Plan for emergencies using an all-hazards approach.
    - Operate effectively in emergency situations.
    - Conduct effective recovery operations after a disaster.

The Role of the Local Emergency Program Manager (Continued)

The Emergency Program Manager coordinates all components of the emergency management system for the community, including:

- Fire and police services.
- Emergency medical programs.
- Public works.
- Volunteers and voluntary organizations.
- Other groups involved in emergency activities.

Other duties of the local Emergency Program Manager might include the following:

- Coordinating the planning process and working cooperatively with organizations and government agencies.
- Advising and informing the Chief Elected Official about emergency management activities.
- Identifying and analyzing the potential effects of hazards that threaten the jurisdiction.
- Taking inventory of personnel and material resources from private sector sources that would be available in an emergency.
- Identifying resource deficiencies and working with appropriate officials on measures to resolve them.
- Developing and carrying out public awareness and education programs.
- Establishing a system to alert officials and the public in an emergency.
- Establishing and maintaining networks of expert advisors and damage assessors for all hazards.
- Coordinating a review of all local emergency-related authorities and recommending amendments, when necessary.

Unit 4: Roles of Key Participants

The Role of the Local Emergency Program Manager (Continued)

Earlier in the course, you reviewed the placement of the emergency manager within local government. Based on the community's organization strategy, the Emergency Program Manager may serve as:

- Part of the fire/rescue department staff.
- Part of a law enforcement agency staff, located in a police department or sheriff's office.
- Head of a separate organization that reports directly to a governing or executive body.

Regardless of location, the person in this position obviously must devote significant time and energy coordinating with a variety of people and organizations within and outside of the community.

Tribal Leader. The tribal leader is responsible for the public safety and welfare of the people of that tribe. As authorized by tribal government, the tribal leader:

Is responsible for coordinating tribal resources needed to prevent, protect against, respond to, and recover from incidents of all types. This also includes preparedness and mitigation activities.

May have powers to amend or suspend certain tribal laws or ordinances associated with response.

Communicates with the tribal community, and helps people, businesses, and organizations cope with the consequences of any type of incident.

Negotiates mutual aid and assistance agreements with other tribes or jurisdictions.

Can request Federal assistance under the Stafford Act through the Governor of the State when it becomes clear that the tribe's capabilities will be insufficient or have been exceeded.

Can elect to deal directly with the Federal Government. Although a State Governor must request a Presidential declaration on behalf of a tribe under the Stafford Act, Federal departments or agencies can work directly with the tribe within existing authorities and resources.

State Emergency Management Role

The role of State government in emergency management in many ways parallels the role of the local emergency management function.

- Legislative and executive authorities exist for State emergency programs, with a range of programs usually operating in a variety of State agencies.

- The State has a responsibility to develop and maintain a comprehensive program for mitigation, preparedness, response, and recovery activities.

The State's role is to supplement and facilitate local efforts before, during, and after emergencies. The State must be prepared to maintain or accelerate services and to provide new services to local governments when local capabilities fall short of demands.

A State government is in the unique position to serve as a link between those who need assistance and those who can assist. It is able to:

- Coordinate with local governments to meet their emergency needs.

- Assess available State and Federal resources.

- Help the local government apply for, acquire, and use those resources effectively.

The State also provides direct guidance and assistance to its local jurisdictions through program development, and it channels Federal guidance and assistance to the local level. In a disaster, the State office helps coordinate and integrate resources and apply them to local needs. The State's role might best be described as "pivotal."

The Governor of a State, who is responsible for the general welfare of the people in that State, has certain legislated powers and resources that can be applied to all-hazards emergency management.

All State Governors have authority and responsibility for:

- Issuing State or area emergency declarations.

- Initiating State response actions (personnel, materials).

- Activating emergency contingency funds and/or reallocating regular budgets for emergency activities.

- Overseeing emergency management for all four phases.

- Applying for and monitoring Federal assistance.

State Emergency Management Role (Continued)

The State Emergency Management Agency:

- Carries out statewide emergency management activities.
- Helps coordinate emergency management activities involving more than one community.
- Assists individual communities when they need help.
- Provides financial assistance on a supplemental basis through a process of application and review.

(The Governor reviews the application, studies the damage estimates and, if appropriate, declares a state of emergency.)

If the local community's resources are not adequate, the first place to turn for additional assistance is to the county or State Emergency Management Agency.

Drawing on these resources occurs during restoration, which involves actions that repair critical infrastructure. This may include restoring utility services, conducting radiological decontamination, and removing debris.

Acting on the information provided, the county or State office will dispatch personnel to the scene to assist in the response and recovery effort. Only the Governor, however, can request the Federal aid that comes with a Presidential Declaration.

State laws require that all States have a State Emergency Management Agency and an EOP coordinated by that agency.

How the Private Sector and Voluntary Organizations Assist Emergency Managers

The private sector, including individuals and families and voluntary organizations, plays a major role in assisting emergency managers before, during, and after an emergency.

- Private industry contributes by:

  - Developing and exercising emergency plans before an emergency occurs.
  - Working with emergency management personnel before an emergency occurs to ascertain what assistance may be necessary and how they can help.
  - Providing assistance (including volunteers) to support emergency management during an emergency and throughout the recovery process.

- Individuals and families contribute by:

  - Taking the time necessary to understand the types of emergencies that are likely to occur and preparing a personal disaster kit and emergency plans for those events.
  - Volunteering with an established organization and receiving training before an emergency occurs.
  - Taking direction and responding reasonably to alerts, warnings, and other emergency public information.

- Voluntary organizations contribute by:

  - Training and managing volunteer resources.
  - Identifying shelter locations and needed supplies.
  - Providing critical emergency services, such as the provision of cleaning supplies, clothing, food, and shelter or assisting with post-emergency cleanup, to those in need.
  - Identifying those whose needs have not been met and coordinating the provision of assistance.

Each of these players is critical to ensuring an appropriate and efficient response. However, each must become involved during the preparedness phase of the integrated emergency management system to ensure that, when an emergency occurs, all players understand their roles and are ready to contribute without delay.

Unit 4: Roles of Key Participants

Federal Emergency Management Role

The Federal government's involvement in emergency management crosses all four phases of emergency management. Assistance may take the form of fiscal support, technical assistance, or information about materials, personnel resources, and research.

The Federal government provides legislation, Executive Orders, and regulations that influence all disaster activities. It also maintains the largest pool of fiscal resources that can be applied to emergency response and recovery.

The Federal Emergency Management Agency (FEMA) takes a lead role in national preparedness for major crises. It also plays coordinating and supportive/assistance roles for integrated emergency management in partnership with State and local emergency management entities. As necessary, FEMA provides funding, technical assistance, services, supplies, equipment, and direct Federal support.

FEMA provides technical and financial assistance to State and local governments to upgrade their communications and warning systems, and it operates an emergency information and coordination center that provides a central location for the collection and management of disaster and emergency information.

FEMA provides information to the President concerning matters of national interest to help with decisions about disaster declarations. The President of the United States is responsible for:

- Protecting the public.

- Making a disaster declaration before Federal funds are released to aid disaster survivors.

The National Response Framework

The NRF is a guide to how the Nation conducts all-hazards incident response. It uses flexible, scalable, and adaptable coordinating structures to align key roles and responsibilities across the Nation. It captures specific authorities and best practices for managing incidents that range from the serious but purely local to large-scale terrorist attacks or catastrophic natural disasters.

The NRF explains the common discipline and structures that have been exercised and have matured at the Local, State, and national levels over time. It captures key lessons learned from Hurricanes Katrina and Rita, focusing particularly on how the Federal government organizes itself to support communities and the States in catastrophic incidents. Most important, it builds upon NIMS, which provides a consistent national template for managing incidents.

The National Response Framework (Continued)

Each ESF is composed of an ESF coordinator and primary/support agencies.

- **ESF Coordinator:** The agency that has ongoing responsibilities throughout the prevention, preparedness, response, recovery, and mitigation phases of incident management for the particular ESF. The ESF coordinating agency is responsible for steady-state planning, preparedness, and other activities.

- **Primary Agencies:** The NRF identifies primary agencies based on authorities, resources, and capabilities.

- **Support Agencies:** Support agencies are assigned based on resources and capabilities in a given functional area. The resources provided by the ESFs reflect the resource typing categories identified in the NIMS.

ESFs may be selectively activated for both Stafford Act and non-Stafford Act incidents. Not all incidents requiring Federal support result in the activation of ESFs.

For Stafford Act incidents, the NRCC or RRCC may activate specific ESFs/or other Federal Agencies (OFA) by directing appropriate departments and agencies to initiate the actions delineated in the ESF Annexes.

Resources coordinated though ESFs are assigned where needed within the response structure. For example, if a State requests assistance with a mass evacuation, resources from several different ESFs may be integrated into a single Branch or Group within the Operations Section. During the response, these resources would report to a supervisor within the assigned Branch or Group.

Note that, regardless of where ESFs may be assigned, they coordinate closely with one another to accomplish their missions.

UNIT 4: ROLES OF KEY PARTICIPANTS

ESF #1 – Transportation
ESF Coordinator: Department of Transportation
- Aviation/airspace management and control
- Transportation safety
- Restoration and recovery of transportation infrastructure
- Movement restrictions
- Damage and impact assessment

ESF #2 – Communications
ESF Coordinator: DHS (National Communications System)
- Coordination with telecommunications and information technology industries
- Restoration and repair of telecommunications infrastructure
- Protection, restoration, and sustainment of national cyber and information technology resources
- Oversight of communications within the Federal incident management and response structures

ESF #3 – Public Works and Engineering
ESF Coordinator: Department of Defense (U.S. Army Corps of Engineers)
- Infrastructure protection and emergency repair
- Infrastructure restoration
- Engineering services and construction management
- Emergency contracting support for lifesaving and lifesustaining services

ESF #4 – Firefighting
ESF Coordinator: Department of Agriculture (U.S. Forest Service)
- Coordination of Federal firefighting activities
- Support to wildland, rural, and urban firefighting operations

ESF #5 – Emergency Management
ESF Coordinator: DHS (FEMA)
- Coordination of incident management and response efforts
- Issuance of mission assignments
- Resource and human capital
- Incident action planning
- Financial management

ESF #6 – Mass Care, Emergency Assistance, Housing, and Human Services
ESF Coordinator: DHS (FEMA)
- Mass care
- Emergency assistance
- Disaster housing
- Human services

ESF #7 – Logistics Management and Resource Support
ESF Coordinators: General Services Administration and DHS (FEMA)
- Comprehensive, national incident logistics planning, management, and sustainment capability
- Resource support (facility space, office equipment and supplies, contracting services, etc.)

Unit 4: Roles of Key Participants

ESF #8 – Public Health and Medical Services
ESF Coordinator: Department of Health and Human Services
- Public health
- Medical
- Mental health services
- Mass fatality management

ESF #9 – Search and Rescue
ESF Coordinator: DHS (FEMA)
- Life-saving assistance
- Search and rescue operations

ESF #10 – Oil and Hazardous Materials Response
ESF Coordinator: Environmental Protection Agency
- Oil and hazardous materials (chemical, biological, radiological, etc.) response
- Environmental short- and long-term cleanup

ESF #11 – Agriculture and Natural Resources
ESF Coordinator: Department of Agriculture
- Nutrition assistance
- Animal and plant disease and pest response
- Food safety and security
- Natural and cultural resources and historic properties protection
- Safety and well-being of household pets

ESF #12 – Energy
ESF Coordinator: Department of Energy
- Energy infrastructure assessment, repair, and restoration
- Energy industry utilities coordination
- Energy forecast

ESF #13 – Public Safety and Security
ESF Coordinator: Department of Justice
- Facility and resource security
- Security planning and technical resource assistance
- Public safety and security support
- Support to access, traffic, and crowd control

ESF #14 – Long-Term Community Recovery
ESF Coordinator: DHS (FEMA)
- Social and economic community impact assessment
- Long-term community recovery assistance to States, tribes, local governments, and the private sector
- Analysis and review of mitigation program implementation

ESF #15 – External Affairs
ESF Coordinator: DHS
- Emergency public information and protective action guidance
- Media and community relations
- Congressional and international affairs
- Tribal and insular affairs

UNIT 4: ROLES OF KEY PARTICIPANTS

Activity: Emergency Management Partners

The purpose of this activity is to match the emergency management partner to a description of that partner's role in emergency management. For a role in Column A, choose a partner from Column B.

Emergency Management Partners	
1. __B__ Declares a local emergency	a. Individual and family ✓
2. __F__ Requests a Presidential Declaration of Disaster	b. Local government official(s) ✓
3. __G__ Activates the National Response Framework	c. Voluntary agency ✓
4. __A__ Responds reasonably to public information	d. Local emergency manager ✓
5. __E__ Supplements and facilitates local emergency efforts	e. State Emergency Management Agency ✓
6. __D__ Coordinates all components of the emergency management system for the community	f. State Governor ✓
	g. FEMA ✓
7. __C__ Provides mass care	

Fundamentals of Emergency Management IS 230.a

UNIT 4: ROLES OF KEY PARTICIPANTS

Activity: Emergency Management Partners (Continued)

Answers to Knowledge Check

1. b
2. f
3. g
4. a
5. e
6. d
7. c

Emergency Management Functional Groups

An integrated approach to emergency management is based on solid general management principles and the common theme of protecting life and property. It provides direction so that participants can begin working together with all of the principals in the network.

On this team are individuals who have obvious responsibilities in emergency response, as well as others whose roles may appear to be minor but which are, in fact, very important. For example, the editor of the local newspaper and the supervisor of a local construction crew may be important members of the emergency management community.

It is helpful to imagine the working relationships of the team as divided into four broadly defined groups at each governmental level, typical of those that exist in many organizations.

- Policy Group. This is an informal and flexible grouping of senior public officials representing State, county, and municipal governments. They meet to develop emergency policies and then, as required by the disaster situation, discuss the economic, political, legal, and social implications of both the threat and the response to determine the best general approach to the situation.

Members of a policy group can include the Governor, Adjutant General, State Director of Emergency Services, County Manager, etc. The Emergency Program Manager serves as the liaison between the policy group and the coordination group.

Unit 4: Roles of Key Participants

Emergency Management Functional Groups (Continued)

- **Coordination Group.** This group typically consists of the assistants, deputies, and staff of agencies and departments represented in the policy group. The coordination group performs a staff function by coordinating the types and number of personnel and material resources deployed, providing logistical support to field units, contracting for relief of forces, and carefully monitoring both the immediate emergency situation and other threats.

  The Emergency Program Manager is responsible for coordinating the efforts of various agency and department personnel assigned to this group. Typically, the coordination group does not command field-level personnel.

- **Operational Response Group.** Persons with specific roles and responsibilities necessary for fully supporting the field response group and who are supervised by an event coordinator are included in this group. The operational response group must include individuals who are responsible for operations, logistics, planning, and finance. These functions interface directly with the ICS structure in the field response group.

- **Field Response Group.** This group represents fire, law enforcement, medical, military, and public works units that normally would be on the scene.

UNIT 4: ROLES OF KEY PARTICIPANTS

 ## Case Study: Emergency Management Coordination

The following description of the response to a tornado illustrates the coordination among local, State, and Federal agencies and the roles of functional groups at different levels. When you have finished reading the case study, answer the questions on the following page.

Tornadoes caused massive property damage and loss of life along a path from northeast Mississippi through central Alabama into northern Georgia. Hardest hit were Jefferson and St. Clair Counties in Alabama. Casualties included 36 fatalities, 273 injuries, and property damage estimated at over $300 million.

After the tornado, the first priority was search and rescue to assist the injured and find missing people. The second priority was to care for people and identify those needing shelter and other assistance.

Local agencies responded quickly to assist the injured and find missing people, joined by county and State agencies coordinated through the Emergency Operations Center (EOC).

The agencies listed below were involved in the response:

Fire and rescue services. Three fire and rescue services from the affected areas coordinated with 26 other fire departments to assist them with search and rescue and emergency medical support.

Law enforcement. The Jefferson County Sheriff's Department and the City of Birmingham police department were the primary responding agencies, assisted by numerous other local and State agencies.

Public Works. The primary response agencies were the Jefferson County Roads and Transportation Department and the City of Birmingham Streets and Sanitation Department, along with Horticulture and Urban Forestry Department, assisted by Public Works Departments from other cities within the county.

EOC Staff. 35 or more agencies, departments, and organizations provided personnel to coordinate the response and recovery efforts.

Other local agencies. Many community groups, churches, State and Federal assistance organizations, and numerous volunteer groups provided assistance, including shelter, mass care, mental health, donated goods and services, and animal rescue and care.

State and Federal agencies. The State Emergency Management Agency coordinated State assistance that included the National Guard, public safety, the State Department of Transportation, and the State Forestry Commission. FEMA provided disaster and community relations teams, established Disaster Recovery Centers to assist survivors seeking individual and housing assistance, and held applicant's briefings for jurisdictions that suffered damage to their infrastructure.

UNIT 4: ROLES OF KEY PARTICIPANTS

Case Study: Emergency Management Coordination (Continued)

Did Alabama receive a Presidential disaster declaration for the disaster? How can you tell?

Yes, FEMA showed up

Name two Jefferson County agencies that were involved in the response and recovery.

Fire/Rescue (EMA)
Law Enforcement (Sheriff)

Name two State agencies or organizations that were involved in the response and recovery.

DDT, AEMA

Name two services performed by voluntary agencies after the tornado.

Shelters
Animal Care

Fundamentals of Emergency Management IS 230.a

Unit 4: Roles of Key Participants

Case Study: Emergency Management Coordination (Continued)

Did Alabama receive a Presidential Declaration of Disaster? How can you tell?

Yes, there was a Presidential Declaration. Disaster Recovery Centers were established, Community Relations and disaster teams were provided, and applicant's briefings for damaged infrastructure were being held.

Name two Jefferson County agencies that were involved in the response and recovery.

Any two of the following:

Sheriff's Department, Emergency Management Agency, Jefferson County Roads and Transportation Department

Name two State agencies or organizations that were involved in the response and recovery.

Any two of the following:

State Emergency Management Agency, National Guard, State Department of Transportation, State Forestry Commission

Name two services performed by voluntary agencies after the tornado.

Any two of the following

Shelter, mass care, mental health services, donated goods and services, and animal rescue and care

Unit 4: Roles of Key Participants

Knowledge Check

Carefully read each question and all of the possible answers before selecting the most appropriate response for each test item. Circle the letter corresponding to the answer you have chosen.

1. The local emergency program manager coordinates:

   a. The National Response Framework.
   b. Search and Rescue.
   c. The National Guard.
   d. The disaster declaration process.
   e. The emergency management programs.

2. The State government is in a unique position to serve as a link between those who need assistance and those who can assist.

   a. True
   b. False

3. If the local community's resources are not adequate to deal with an emergency, it can request assistance from:

   a. Neighboring jurisdictions.
   b. Congressional representatives.
   c. The President.
   d. FEMA.

4. If a disaster response demands more resources than any local governments can supply without assistance, the next step is to request assistance from:

   a. Congressional representatives.
   b. The President.
   c. FEMA.
   d. The State Emergency Management Agency.

Fundamentals of Emergency Management IS 230.a

UNIT 4: ROLES OF KEY PARTICIPANTS

Knowledge Check (Continued)

Answers to the Knowledge Check

1. e
2. a
3. a
4. d

Unit 5: The Plan as a Program Centerpiece

Unit 5: The Plan as a Program Centerpiece

Introduction and Unit Overview

This unit will describe the hazard analysis process and why conducting a thorough hazard analysis forms the basis for all emergency operations planning. After you complete this unit, you should be able to:

- Describe why an Emergency Operations Plan (EOP) must be the centerpiece of an Emergency Management program.
- Identify the high-risk hazards facing their communities.
- Describe the structure of an EOP.
- Determine the annexes and appendixes that should be included in their plans.

What is an EOP and What Does It Do?

An EOP is a key component of an emergency operations program. It establishes the overall authority, roles, and functions performed during emergencies.

The emergency operations program is in continuous operation and includes nonemergency activities such as training and exercises. The EOP does not come into play for nonemergency activities.

The EOP is activated to guide emergency response and recovery, and is implemented only after being triggered by emergencies. Even during emergencies, only those parts of the EOP that are required for the response are activated.

What is an EOP and What Does It Do? (Continued)

The EOP enables the community to be prepared to take immediate action when disaster threatens or strikes. An EOP describes:

- What emergency response actions will occur. . .
- Under what circumstances. . .
- By whom. . .
- By what authority. . .
- Using what resources.

An EOP describes how the government does business in an emergency.

Developing and maintaining an EOP is a crucial task of an emergency operations program. The program should bring together representatives of all involved agencies who could have a role in a response. Agency participation is necessary to ensure that all who have a role in a response are brought into the process and understand their responsibilities thoroughly.

To illustrate how an EOP organizes a response to an emergency, consider the train derailment and subsequent hazardous chemical release described earlier in the course.

UNIT 5: THE PLAN AS A PROGRAM CENTERPIECE

Activity: Where Do I Fit Into the EOP?

This activity will provide you with an opportunity to analyze the role you (and your agency) would play in a response similar to the train derailment. To complete this activity, respond to the questions below. You may need to refer to the local EOP for this activity.

1. Would your agency or organization participate in the response described in the case study?

    ☐ Yes
    ☐ No

    If your agency would not be deployed based on the procedures in your community's EOP, do you see a role that it might have played? If yes, describe the role that your agency could perform.

2. If you answered yes to question 1, where would your agency deploy?

    ☐ To the scene
    ☐ To the EOC
    ☐ Representatives would be deployed to the scene and to the EOC

3. Describe the specific duties your agency would perform in the case study.

Unit 5: The Plan as a Program Centerpiece

 ## Case Study: An EOP in Action

The freight train derailed at approximately 3:00 AM, releasing anhydrous ammonia.

The local EOP designates the Emergency Management Agency (EMA) as the county's 24-hour crisis monitor. The EMA received first notification of the incident from a railroad official shortly after 3:00 A.M.

Because the derailment occurred outside of the city limits, the Chairman of the County Board of Supervisors had direction and control under the plan, and declared a State of Emergency in the county soon after being notified.

Each department listed in the plan was notified and alerted its employees and volunteers. The EMA activated the Emergency Operations Center (EOC). County communications staff moved operations to the EOC, and following the procedures in the Communications Annex, representatives of fire departments, law enforcement agencies, public works agencies, and the Department of Health deployed to the EOC.

The plan designated the hazardous chemical release as a level 3 emergency, which is used for all major technological disasters. The State EOC was notified, as was required for level 3 emergencies.

The designated on-scene Incident Commander was the County Fire Chief. The fire department checked the Material Safety Data Sheet (MSDS) for anhydrous ammonia and discovered that protective gear would be needed for any response personnel at the scene. No one was allowed near the toxic gas cloud until gear could be obtained.

A Hazardous Materials Appendix to the EOP listed sources for protective gear, protective actions that could be taken, and information that should be given to the public. The appendix established cleanup of the site as the responsibility of the railroad company in coordination with the county fire department, as the controlling authority for incidents involving hazardous materials.

The EMA followed the procedures included in the Warning Annex by activating warning sirens and broadcasting instructions to "shelter in place" by closing all windows and turning off furnaces to avoid bringing outside air into their homes. Public works employees set up a perimeter a safe distance from the scene that was manned by police and sheriff's officers to limit access to the release area.

After the immediate danger passed, the "shelter in place" advisory was lifted, and survivors could seek medical treatment. Hospitals activated procedures to mobilize extra staff to treat hundreds of survivors suffering from exposure to the chemical.

Unit 5: The Plan as a Program Centerpiece

Importance of the Hazard Analysis to the Planning Process

When completed, the community's hazard analysis should form the basis for the entire emergency planning process because it will guide response actions by highlighting:

- The hazards that pose the greatest risks to the community.
- The types and degree of damage the can be expected for each type of hazard, including the areas and populations with the highest probability for damage.
- The types of resources that will most likely be needed to respond.
- Potential resource shortfalls that need to be filled.

Refer to the hazard analysis throughout the planning process to help keep the planning team on track.

What Is In a Hazard Analysis?

As described in the previous unit, a hazard analysis involves examining the likely hazards that could affect your community and quantifying the risk posed by each hazard.

Hazards are conditions or situations that have the potential for causing harm to people, property, or the environment.

Hazards can be classified into three categories:

- Natural (e.g., tornadoes and earthquakes)
- Intentional (e.g., terrorism or civil disturbance)
- Technological (e.g., failure of the power grid or hazardous materials spills)

Hazard analysis itself includes three steps:

1. Identifying the types of hazards to which a community is vulnerable.
2. Developing a profile of each hazard.
3. Quantifying the risk posed by each hazard.

Job Aid 5.1 on the following page is a sample Hazard Analysis Worksheet. Developing a worksheet such as that shown in the job aid can help you determine how high a risk your community faces and what your resulting response needs may be.

UNIT 5: THE PLAN AS A PROGRAM CENTERPIECE

Job Aid 5.1: Hazard Analysis Worksheet

Hazard: _____

Frequency of Occurrence:

- ☐ *Highly likely* (Near 100% probability in the next year)
- ☐ *Likely* (Between 10% and 100% probability in the next year, or at least one chance in the next 10 years)
- ☐ *Possible* (Between 1% and 10% probability in the next year, or at least one chance in the next 100 years)
- ☐ *Unlikely* (Less than 1% probability in the next 100 years)

Seasonal pattern?

- ☐ No
- ☐ Yes. Specify season(s) when hazard occurs: _____

Potential Impact:

- ☐ *Catastrophic* (Multiple deaths; shutdown of critical facilities for 1 month or more; more than 50% of property severely damaged)
- ☐ *Critical* (Injuries or illness resulting in permanent disability; shutdown of critical facilities for at least 2 weeks; 25% to 50% of property severely damaged)
- ☐ *Limited* (Temporary injuries; shutdown of critical facilities for 1-2 weeks; 10% to 25% of property severely damaged)
- ☐ *Negligible* (Injuries treatable with first aid; shutdown of critical facilities for 24 hours or less; less than 10% of property severely damaged)

Are any areas or facilities more likely to be affected (e.g., air, water, or land; infrastructure)? If so, which?

Fundamentals of Emergency Management IS 230.a

Job Aid 5.1: Hazard Analysis Worksheet (Continued)

Speed of Onset:

☐ *Minimal or no warning*
☐ *6 to 12 hours warning*
☐ *12 to 24 hours warning*
☐ *More than 24 hours warning*

Potential for Cascading Effects?

☐ No
☐ Yes. Specify effects:

Unit 5: The Plan as a Program Centerpiece

Hazard Identification

The first step, then, is to develop a list of hazards that may occur in your community. This list is usually based on historical data about past events. Sources of information about past events may include:

- Newspaper files.
- Weather records.
- Insurance records.
- Accident reports (e.g., for hazardous materials incidents).
- EOC records.
- Fire department inspections.
- Anecdotal information from long-time residents.

Be aware that information about recent or costly events is relatively easy to locate, but information about older or less costly events may be more difficult to find. Do not ignore hazards simply because they have not occurred recently or, at last occurrence did not cause extensive damage. A hazard that was low impact 15 years ago may have catastrophic consequences today because of the hazard itself or because of changes (e.g., new development in higher-risk areas) since the last occurrence.

Keep in mind that you may need to consider potential hazards that are located in neighboring jurisdictions if the hazards could pose a threat to your community or affect your community's response. For example, an explosion in a hazardous materials plant on the border of a neighboring State could result in toxic fumes in your community, depending on the wind direction; a dam that breaks in a city located upstream could result in a flash flood in your community.

Unit 5: The Plan as a Program Centerpiece

Developing Hazard Profiles

Merely identifying the hazards that occur in your community does not tell you how to plan for those hazards. The next step is to develop a profile of each identified hazard according to the following characteristics:

- **Predictability** (i.e., frequency and/or likelihood of occurrence, seasonal pattern, etc.).

- **Magnitude** or severity of impact on the community (i.e., extent of damage expected, the types of damage that can be expected to the infrastructure, etc.).

- **Speed of onset** (e.g., hurricanes usually provide some amount of preparation time before they strike, while earthquakes or explosions could occur without warning).

- The potential for **cascading effects** (e.g., flooding following a hurricane or fires following an earthquake because of gas line ruptures).

Using a Hazard Analysis to Determine Risk

After compiling information for each hazard that your community is vulnerable to, your next step is to assess the risks associated with each hazard so that the planning team can predict and prepare for those of highest potential impact on people, services, facilities and structures.

When assessing risk, it is important to keep in mind the following hierarchy of response priorities:

1. **Life safety.** Conditions that could affect the health and/or safety of the population.

2. **Essential facilities.** Facilities, such as fire houses, precinct houses or waste water treatment facilities that, if affected by the hazard would seriously and adversely affect the community's ability to respond.

3. **Critical infrastructure.** Roadways, utilities, and other components of the infrastructure that, if damaged, would seriously and adversely affect life safety or response capability.

When surveying your community for risks to these important resources, include such elements as geographic features, infrastructure lifelines, essential facilities, special facilities, population densities and shifts (demographics), and availability of response resources.

Unit 5: The Plan as a Program Centerpiece

EOP and Developing and Maintaining State and Territorial Local Government Emergency Plans (Comprehensive Planning Guidance 101)

The NRF identifies State, Territorial, tribal, and Local jurisdiction responsibility to develop detailed, robust all-hazards/all-threats EOPs. It says these plans:

- Should clearly define leadership roles and responsibilities and clearly articulate the decisions that need to be made, who will make them, and when;
- Should include both hazard- and threat-specific and all-hazards/all-threats plans tailored to the locale;
- Should be integrated and operational and incorporate key private sector business and NGO elements; and
- Should include strategies for both no-notice and forewarned evacuations, with particular consideration given to assisting special-needs populations.

CPG 101 is the foundation for State and local planning in the United States. CPG 101 expands on the Federal Emergency Management Agency's (FEMA's) efforts to provide guidance about response and recovery planning to State, Territorial, tribal, and Local governments. It also extends those planning concepts into the prevention and protection mission areas. Some predecessor material can be traced back to the 1960s-era *Federal Civil Defense Guide*. Long-time emergency management practitioners also will recognize the influence of Civil Preparedness Guide 1-8, *Guide for the Development of State and Local Emergency Operations Plans*, and State and Local Guide (SLG) 101, *Guide for All-Hazards Emergency Operations Planning*, in this document.

CPG 101 provides general guidelines on developing emergency operations plans. It promotes a common understanding of the fundamentals of planning and decision making to help operations planners examine a hazard or threat and produce integrated, coordinated, and synchronized plans. This Guide helps emergency and homeland security managers in State, Territorial, tribal, and Local governments (hereafter, State and Local governments) in their efforts to develop and maintain viable all-hazard, all-threat emergency plans. Each jurisdiction's plans must reflect what *that* **community** will do to protect itself from *its* unique hazards and threats with the unique resources *it* has or can obtain.

While CPG 101 maintains its link to the past, it also reflects the changed reality of the current operational planning environment. Hurricane Hugo and the Loma Prieta earthquake influenced the development of CPG 1-8. Hurricane Andrew and the Midwest floods shaped the contents of SLG 101. In a similar way, CPG 101 reflects the impacts of the September 11, 2001, terrorist attacks and recent major disasters, such as Hurricanes Katrina and Rita, on the emergency planning community. CPG 101 integrates concepts from the National Preparedness Guidelines (NPG), National Incident Management System (NIMS), National Response Framework (NRF), National Strategy for Information Sharing (NSIS), and National Infrastructure Protection Plan (NIPP), and it incorporates recommendations from the 2005 Nationwide Plan Review. CPG 101 also serves as a companion document to the Integrated Planning System (IPS) mandated by Annex I of Homeland Security Presidential Directive (HSPD)-8, and it fulfills the requirement that the IPS address State and local planning.

Unit 5: The Plan as a Program Centerpiece

Additionally, CPG 101 also references the Target Capabilities List (TCL) that outlines the fundamental capabilities essential to implementing the *National Preparedness Guidelines*. As part of a larger planning modernization effort, CPG 101 provides methods for State, Territorial, tribal, and Local planners to:

- Develop sufficiently trained planners to meet and sustain planning requirements;
- Identify resource demands and operational options across all homeland security mission areas throughout the planning process;
- Link planning, preparedness, and resource and asset management processes and data in a virtual environment;
- Prioritize plans and planning efforts to best support emergency management and homeland security strategies and allow for their seamless transition to execution;
- Produce and tailor the full range of combined Federal, State, Territorial, tribal, and Local government options according to changing circumstances; and
- Quickly produce plans on demand, with revisions as needed

CPG 101 provides emergency and homeland security managers and other emergency services personnel with FEMA's recommendations on how to address the entire planning process — from forming a planning team, through writing and maintaining the plan, to executing the plan. It also encourages emergency and homeland security managers to follow a process that addresses all of the hazards and threats that might impact their jurisdiction through a suite of operations plans (OPLANs) connected to a single, integrated concept plan (CONPLAN).

Planning includes *senior officials* throughout the process to ensure both understanding and buy-in. Potential planning team members have many day-to-day concerns. A planning team's members must be convinced that emergency planning is a high priority. Chief executive support helps the planning process meet requirements of time, planning horizons, simplicity, and level of detail. The more involved decision makers are in planning, the better the planning product will be.

Unit 5: The Plan as a Program Centerpiece

PLANNING PRINCIPLES

The challenge of developing an all-hazards plan for protecting lives, property, and the environment is made easier if the planners preparing it apply the following principles to the planning process:

Planning must involve *all* partners. Just as coordinated emergency operations depend on teamwork, good planning requires a team effort. The most realistic and complete plans are prepared by a team that includes representatives of the departments and agencies, as well as the private sector and NGOs that can contribute critical perspectives or that will have a role in executing the plan. This principle is so important that the first step of the planning process is forming a planning team. When the plan considers and incorporates the views of the individuals and organizations assigned tasks within it, they are more likely to accept and use the plan.

Emergency operations planning addresses all hazards and threats. The causes of emergencies can vary greatly, but many of the effects do not. Planners can address common operational functions in the basic plan instead of having unique plans for every type of hazard or threat. For example, floods, wildfires, hazardous materials releases, and radiological dispersion devices (RDDs) may lead a jurisdiction to issue an evacuation order and open shelters. Even though each hazard's characteristics (e.g., speed of onset, size of the affected area) are different, the general tasks for conducting an evacuation and shelter operations are the same. While differences in the speed of onset may affect when the order to evacuate or to open and operate shelters is given, the process of determining the need for evacuation or shelters and issuing the order does not change. All hazards and all-threats planning ensure that, when addressing emergency functions, planners identify common tasks and who is responsible for accomplishing those tasks.

Planning does not need to start from scratch. Planners should take advantage of existing plans and others' experience. The State is a valuable resource for the Local jurisdiction. Many States publish their own standards and guidance for emergency planning, conduct workshops and training courses, and assign their planners to work with local planners. FEMA supports State training efforts through its National Preparedness Directorate by offering resident, locally presented, and independent-study emergency planning courses. FEMA also publishes many documents related to planning for specific functions and hazards and threats. By reviewing existing emergency or contingency plans, planners can:

- Identify applicable authorities and statutes;
- Gain insight into community risk perceptions;
- Identify organizational arrangements used in the past;
- Identify mutual aid agreements with other jurisdictions;
- Identify private sector planning that can complement and focus public sector planning;
- Learn how some planning issues were resolved in the past; and
- Identify preparedness gaps in available personnel, equipment, and training.

Unit 5: The Plan as a Program Centerpiece

The emergency or homeland security manager should seek the chief executive's support for and involvement in the planning effort. The emergency or homeland security manager must explain to the chief executive what is at stake in emergency planning by:

- Identifying and sharing the hazard, risk, and threat analyses for the jurisdiction;
- Describing what the government body and the chief executive will have to do prior to, during, and after an event;
- Determining what the government body and the chief executive can do to either prevent or minimize the impact of an event;
- Discussing readiness assessments and exercise critiques; and
- Reaffirming the chief executive's understanding that planning is an iterative, dynamic process that ultimately facilitates his or her job in a crisis.

Planning is influenced by time, uncertainty, risk, and experience. These factors define the starting point where planners apply appropriate concepts and methods to create solutions to particular problems. Because this activity involves judgment and the balancing of competing demands, plans cannot be overly detailed — to be followed to the letter — or so general that they provide insufficient direction. This is why planning is both science and art, and why plans are evolving frameworks.

STEPS IN THE PLANNING PROCESS

There are many ways to produce an operations plan. The planning process that follows has enough flexibility for each community to adapt it to its unique characteristics and situation. Small communities can follow just the steps that are appropriate to their size, known hazards and threats, and available planning resources. The steps of this process are to:

1. Form a collaborative planning team.
2. Understand the situation.
    (a) Conduct research.
    (b) Analyze the information.
3. Determine goals and objectives.
4. Develop the plan.
    (a) Develop and analyze courses of action.
    (b) Identify resources.
5. Prepare, review, and gain approval of the plan.
    (a) Write the plan.
    (b) Approve and disseminate the plan.
6. Refine and execute the plan.
    (a) Exercise the plan and evaluate its effectiveness.
    (b) Review, revise, and maintain the plan.

Potential Members of a Larger Community Planning Team

Individuals/Organizations	What They Bring to the Planning Team
Senior Official (elected or appointed) or designee	<ul><li>Support for the homeland security planning process</li><li>Government intent by identifying planning goals and essential tasks</li><li>Policy guidance and decision-making capability</li><li>Authority to commit the jurisdiction's resources</li></ul>
Emergency Manager or designee	<ul><li>Knowledge about all-hazard planning techniques</li><li>Knowledge about the interaction of the tactical, operational, and strategic response levels</li><li>Knowledge about the prevention, protection, response, recovery, and mitigation strategies for the jurisdiction</li><li>Knowledge about existing mitigation, emergency, continuity, and recovery plans</li></ul>
EMS Director or designee	<ul><li>Knowledge about emergency medical treatment requirements for a variety of situations</li><li>Knowledge about treatment facility capabilities</li><li>Specialized personnel and equipment resources</li><li>Knowledge about how EMS interacts with the Emergency Operations Center and incident command</li></ul>
Fire Services Chief or designee	<ul><li>Knowledge about fire department procedures, on-scene safety requirements, hazardous materials response requirements, and search-and-rescue techniques</li><li>Knowledge about the jurisdiction's fire-related risks</li><li>Specialized personnel and equipment resources</li></ul>
Law Enforcement Chief or designee	<ul><li>Knowledge about police department procedures; on-scene safety requirements; local laws and ordinances; explosive ordnance disposal methods; and specialized response requirements, such as perimeter control and evacuation procedures</li><li>Knowledge about the prevention and protection strategies for the jurisdiction</li><li>Knowledge about fusion centers and intelligence and security strategies for the jurisdiction</li><li>Specialized personnel and equipment resources</li></ul>
Public Works Director or designee	<ul><li>Knowledge about the jurisdiction's road and utility infrastructure</li><li>Specialized personnel and equipment resources</li></ul>
Public Health Officer or designee	<ul><li>Records of morbidity and mortality</li><li>Knowledge about the jurisdiction's surge capacity</li><li>Understanding of the special medical needs of the community</li><li>Knowledge about historic infectious disease and syndromic surveillance</li><li>Knowledge about infectious disease sampling procedures</li></ul>
Hazardous Materials Coordinator	<ul><li>Knowledge about hazardous materials that are produced, stored, or transported in or through the community</li><li>Knowledge about U.S. Environmental Protection Agency (EPA), Occupational Safety and Health Administration (OSHA), and U.S. Department of Transportation (DOT) requirements for producing,</li></ul>

		storing, and transporting hazardous materials and responding to hazardous materials incidents
Hazard Mitigation Specialist		<ul><li>Knowledge about all-hazard planning techniques</li><li>Knowledge of current and proposed mitigation strategies</li><li>Knowledge of available mitigation funding</li><li>Knowledge of existing mitigation plans</li></ul>
Transportation Director or designee		<ul><li>Knowledge about the jurisdiction's road infrastructure</li><li>Knowledge about the area's transportation resources</li><li>Familiarity with the key local transportation providers</li><li>Specialized personnel resources</li></ul>
Agriculture Extension Service		<ul><li>Knowledge about the area's agricultural sector and associated risks (e.g., fertilizer storage, hay and grain storage, fertilizer and/or excrement runoff)</li></ul>
School Superintendent or designee		<ul><li>Knowledge about school facilities</li><li>Knowledge about the hazards that directly affect schools</li><li>Specialized personnel and equipment resources (e.g., buses)</li></ul>
Social services agency representatives		<ul><li>Knowledge about special-needs populations</li></ul>
Local Federal asset representatives		<ul><li>Knowledge about specialized personnel and equipment resources that could be used in an emergency</li><li>Facility security and response plans (to be integrated with the jurisdiction's EOP)</li><li>Knowledge about potential threats to or hazards at Federal facilities (e.g., research laboratories, military installations)</li></ul>
NGOs (includes members of National VOAD [Voluntary Organizations Active in Disaster]) and other private, not-for-profit, faith-based, and community organizations		<ul><li>Knowledge about specialized resources that can be brought to bear in an emergency</li><li>Lists of shelters, feeding centers, and distribution centers</li><li>Knowledge about special-needs populations</li></ul>
Local business and industry representatives		<ul><li>Knowledge about hazardous materials that are produced, stored, and/or transported in or through the community</li><li>Facility response plans (to be integrated with the jurisdiction's EOP)</li><li>Knowledge about specialized facilities, personnel, and equipment resources that could be used in an emergency</li></ul>
Amateur Radio Emergency Service (ARES)/Radio Amateur Civil Emergency Services (RACES) Coordinator		<ul><li>List of ARES/RACES resources that can be used in an emergency</li></ul>
Utility representatives		<ul><li>Knowledge about utility infrastructures</li><li>Knowledge about specialized personnel and equipment resources that could be used in an emergency</li></ul>
Veterinarians/animal shelter representatives		<ul><li>Knowledge about the special response needs for animals, including livestock</li></ul>

Unit 5: The Plan as a Program Centerpiece

This format is not required but is based on FEMA's experience in responding to disasters and with working with State and local governments as they develop their plans. Regardless of how your community's EOP is organized, the important point to note is that it must be easy to understand and easy to use *by all* who have a role in a response.

The Basic Plan

The Basic Plan provides an overview of your community's response organization and policies. It also cites the legal authority for conducting emergency operations, describes the hazards that the EOP is intended to address, explains the general concept of emergency operations, and assigns responsibility for emergency planning and operations.

Unit 5: The Plan as a Program Centerpiece

The Basic Plan (Continued)

The Basic Plan is typically organized into the following sections:

- **Introductory Material.** The introductory material provides the authority and responsibility for responding agencies to perform their tasks under the plan. It also facilitates the ease of use for the overall document. Typically, the introductory material includes:

    - A promulgation document that provides the legal authority and the responsibility to respond to emergencies.
    - A signature page that includes the signatures of the agency executives for responding agencies. The signature page indicates that the signatory agencies have worked together in the plan's development and agree to the performance commitments made in the plan.
    - A dated title page and record of changes that indicates the date of original publication and of any subsequent changes to the plan. Including a change record in the Basic Plan helps users keep the plan up to date and know that they are using the most recent version.
    - A record of distribution that indicates the individuals and agencies (or organizations) that received a copy of the plan. The record of distribution provides proof that the EOP has been distributed and that the individuals and agencies have had a chance to review the plan.
    - A Table of Contents that includes all of the section titles and subtitles for the plan to provide a topical overview of the document.

- **Purpose.** The Purpose statement explains why the plan has been developed and what it is meant to do. When properly developed, all other information contained in the plan flows logically from the purpose statement.

- **Situation and Assumptions.** The Situation and Assumptions statement provides a statement of the scope of the EOP, outlining the hazards that the plan addresses, community characteristics that may affect the response, and assumptions on which the plan is based (e.g., that, in the case of a catastrophic disaster affecting adjacent communities, mutual aid might not be available).

The Basic Plan (Continued)

- <u>Concept of Operations</u>. The Concept of Operations provides a basic statement of *what response activities should occur, within what timeframe,* and *at whose direction.* A good Concept of Operations describes the community's approach to emergency response. Typically, the Concept of Operations should include such topics as:

    - Division of responsibilities between local and State responders.
    - The procedure for activating the EOP.
    - Alert levels and the tasks that should be performed at each level.
    - The general sequence of actions to be taken before, during, and after an emergency.
    - Who can request aid and under what conditions.

- <u>Organization and Assignment of Responsibilities</u>. The Organization and Assignment of Responsibilities describes how the community will be organized to respond to emergencies. The section includes a list, by position and organization, of the types of tasks that will be performed. At a minimum, the Organization and Assignment of Responsibilities should include a task listing for the:

    - Chief Elected Official.
    - Fire Department.
    - Police Department.
    - Health and Medical Coordinator.
    - Public Works Department.
    - Warning Coordinator.
    - EOC Manager.
    - Emergency Manager.
    - Communications Coordinator.
    - Public Information Officer (PIO).
    - Evacuation Coordinator.
    - Mass Care Coordinator.
    - Resource Manager.
    - School Superintendent.
    - Animal Care and Control Agency.

The Basic Plan (Continued)

- **Administration and Logistics.** The Administration and Logistics section describes the support requirements and the availability of support and services for all types of emergencies. It also includes general policies for managing resources, including policies on keeping financial records, reporting, tracking resource needs, tracking the source and use of resources, procurement, and compensating owners of private property used by the community during the response. Mutual aid agreements with neighboring jurisdictions should be referenced, but not included, in this section.

- **Plan Development and Maintenance.** The Plan Development and Maintenance section describes the community's overall approach to planning, including the assignment of planning responsibilities.

- **Authorities and References.** The Authorities and References section should provide the legal basis for emergency operations. The section should include a list of laws, statutes, ordinances; Executive Orders, regulations, and formal agreements related to emergency response. This section should also provide the limits of the emergency authority of the Chief Elected Official, the circumstances under which the authorities become effective, and when they are terminated.

The Basic Plan may also include maps of the community and other documents that will assist the overall response. Despite the number of sections in the Basic Plan, it need not be long and complicated. In fact, a simple, concise Basic Plan that is easy to use is far preferable to one that includes too many details and too much verbiage.

Functional Annexes

Functional annexes include those parts of the plan that are organized around broad functions. For example, evacuation and communication are functions that are typically included in annexes. Each annex focuses on one function that the community believes will be necessary during an emergency. The number and type of functional annexes may vary, depending on the community's needs, capabilities, risks, and resources.

Functional Annexes (Continued)

FEMA recommends that communities include the functions listed below as functional annexes to their Basic Plan:

- Direction and Control. This annex allows the community to analyze the emergency and decide how to respond by directing and coordinating the efforts of the jurisdiction's response forces and coordinating with the mutual aid partners to use all resources efficiently and effectively.

- Communications. This annex focuses on the systems that will be relied on for responders and other emergency personnel to communicate with each other (i.e., not with the public) during emergencies. It describes the total communications system, including backup systems, and provides procedures for its use.

- Warning. This annex describes the warning systems in place and the responsibilities and procedures for issuing warnings to the public. All components of the warning system should be described, including contingency plans, to ensure that warning information is available to the public.

- Emergency Public Information. The Emergency Public Information (EPI) Annex describes the methods that the community will use to provide information to the public before, during, and after an emergency. Historically, the EPI Annex has been developed based on the assumption that an emergency is imminent or has occurred. Recent terrorism incidents, however, demand that some degree of preparedness be incorporated into the EPI Annex so that, when a terrorist incident occurs, the public is already aware of the potential implications of the incident and understands that government authorities are doing everything possible to control the situation. This expansion of the concept behind EPI will help ensure that the public takes the appropriate action. It will also minimize a panic response among the public and will give the public confidence that the government is in control.

- Evacuation. The Evacuation Annex addresses the movement of people from an area that has been affected by an emergency to a safe area. Considerations for evacuating persons with special needs should always be included in the Evacuation Annex.

Unit 5: The Plan as a Program Centerpiece

Functional Annexes (Continued)

- **Mass Care**. This annex addresses the actions that will be taken to protect evacuees and others from the effects of the event. The Mass Care Annex describes how sheltering, food distribution, medical care, clothing, and other essential life support needs will be provided to those who have been displaced by a hazard. (Note that communities that are at risk from hurricanes should include a discussion of refuges of last resort in this annex.)

- **Health and Medical**. The Health and Medical annex addresses the activities associated with the provision of health and medical services in emergencies, including emergency medical, hospital, public health, environmental health, mental health, and mortuary services.

- **Resource Management**. Because emergencies can require more—and more specialized—resources than responding agencies have available, the Resource Management Annex facilitates the identification of existing resources, the identification of probable resource needs, and a description of how additional resources will be acquired and distributed.

Annexes should be organized in the same way as the Basic Plan (i.e., Purpose, Situation and Assumptions, etc.) but should not repeat the information that is included in the Basic Plan. Rather, annexes should include only the information that is specific to the emergency function covered by the annex.

Hazard-Specific Appendixes

Appendixes to plan annexes should be developed for each hazard that the plan addresses (e.g., tornado, earthquake, terrorism), and the decision of whether or not to include a specific annex should be based on the community's hazard analysis. They are developed based on special planning requirements that are not common across all hazards addressed by an annex.

By developing hazard-specific appendixes, planners address the special or unique response considerations related to each hazard for which the community is at high risk, including regulatory requirements associated with specific hazard types (e.g., hazardous materials). Appendixes are supplements to functional annexes.

Like annexes, appendixes should be organized in the same way as the Basic Plan and should not repeat information that is included in either the Basic Plan or the annexes to which they are attached.

Implementing Instructions

Implementing instructions delineate the actual procedures that response personnel will follow in an emergency. Although many local and State response agencies refer to implementing instructions as Standard Operating Procedures (SOPs), they are actually much more than SOPs. Implementing instructions include any resource that responders may use to help them remember what to do in an emergency. In addition to SOPs, implementing instructions may include:

- Checklists.
- Worksheets.
- Instruction cards.

Implementing instructions are developed by the response agencies and are included in the EOP by reference only.

UNIT 5: THE PLAN AS A PROGRAM CENTERPIECE

Knowledge Check

Carefully read each question and all of the possible answers before selecting the most appropriate response for each test item. Circle the letter corresponding to the answer you have chosen.

1. A hazard analysis is the first step in the emergency planning process.

   a. True
   b. False

2. The first step in a hazard analysis is _____.

   a. Identifying sources of additional resources needed to respond to each hazard.
   b. Determining the risk that each hazard poses to your community.
   c. Prioritizing the risk of each hazard to your community.
   d. Identifying all hazards that pose a risk to your community.

3. A hazard that causes a second emergency event to occur is called a:

   a. Double event.
   b. Cascading event.
   c. Major disaster.
   d. Complex response.

4. When assessing risk, the top response priority is:

   a. Essential facilities.
   b. Critical Infrastructure.
   c. Life safety.

5. The Organization and Assignment of Responsibilities describes how the community will be organized to respond to emergencies.

   a. True
   b. False

Unit 5: The Plan as a Program Centerpiece

Knowledge Check (Continued)

1. a
2. d
3. b
4. c
5. a

Unit 6: Planning and Coordination

Unit 6: Planning and Coordination

Introduction and Unit Overview

An integrated emergency management system depends on an EOP to organize response and other emergency activities. Most communities also have other plans, such as a comprehensive plan, that should be integrated with emergency management. In this unit, you will learn about:

- Identifying resources that your organizations can offer during an emergency.

- Describing the benefits of using the Incident Command System for emergency response.

- Describing the interrelationships between ICS and the Emergency Operations Center.

- Listing four ways to augment local resources, and give an example of when each is appropriate.

Linking Hazard Analysis to Capability Assessment

Each responding agency or organization in the community should have personnel rosters, training records, equipment inventories, and other information needed to develop a complete picture of the resources that are available in an emergency. During the planning process, this information should be compared with anticipated resource needs for emergencies of varying types and scales. Only by completing such a comparison can resource shortfalls be anticipated.

It would be useful to maintain a list of resources needed to respond to all or most of your community's identified hazards. The list should include:

- The resource type and number available.

- The number of each item that is available.

- A point of contact and 24-hour contact phone numbers for activation of the resource.

- The cost or fee for use of the resource.

- The date that resource availability was last verified.

- Procedures for inspection, pick-up, and return of the resource.

After identifying potential resource shortfalls, the planning team must determine how to obtain the additional resources necessary. Some options for obtaining needed resources include:

- Mutual Aid and Assistance Agreements with neighboring jurisdictions. Mutual aid agreements are usually voluntary agreements to pool resources when any participating community experiences a shortfall. The most common agreements are for fire, police, and Emergency Medical Services (EMS) services, but these agreements can be developed for any type of resource.

    Note: Do not assume that, because your community has agreements in place, its resource needs will be met. Emergencies that affect multiple communities may make resource sharing impossible. Contingency plans should be developed to deal with situations in which mutual aid is not available.

- Standby contracts are contracts for critical equipment and supplies that become effective only if necessary following an emergency event. Typically, standby contracts establish prices as those in effect on the day before the event occurred. The use of standby contracts can help ensure that emergency supplies are available in the quantities needed and at a reasonable price.

Linking Hazard Analysis to Capability Assessment (Continued)

- Private-sector organizations that have specialized expertise and equipment. Often, industrial facilities have their own response personnel and equipment that can be called upon in a general emergency.

- Local military installations have a sense of ownership in the community. They also have personnel with specialized training and equipment that can be used in a general emergency. Local governments can develop agreements, similar to mutual aid and assistance agreements, with commanders of local military installations to augment local response capabilities.

- State governments have additional technical and response capabilities that can be requested when local resources are over stretched. Additionally, most State governments have Emergency Mutual Aid Compacts (EMACs), which are similar to the local mutual aid and assistance agreements, with neighboring States to supplement their resources.

- The Federal government can provide technical and other emergency assistance when requested by the Governor of the affected State *if* the President declares the area a major emergency or disaster. When an emergency or disaster declaration occurs, Federal departments and agencies provide full and prompt cooperation, available resources, and support consistent with their authorities.

There may be other ways to obtain the resources necessary for a given response need. Think expansively as you consider resource management so that no potential resource is missed.

UNIT 6: PLANNING AND COORDINATION

Activity: What Can Your Organization Offer?

This activity is designed to make you think about the types of resources (i.e., personnel, tools, and equipment) that your organization or agency could provide in the event of an emergency. Consider your agency's role in an emergency response. Then, list the resources that the agency could provide to fulfill its response requirements.

Agency: _____

Primary Response Role: _____

Personnel	Tools	Equipment

Fundamentals of Emergency Management

The EOP and the Incident Command System

An EOP calls for a coordinated response to various events from a number of different governmental, private sector, and volunteer organizations.

Unit 2 described the NIMS requirement to institutionalize ICS, as well as how ICS fits with the other NIMS standard incident command structures and the major components of NIMS.

As defined in NIMS, ICS:

- Is a standardized, on-scene, all-hazard incident management concept.
- Allows its users to adopt an integrated organizational structure to match the complexities and demands of single or multiple incidents without being hindered by jurisdictional boundaries.

ICS is an emergency management model for command and management and coordination of a response operation. It is based on core principles that have been proven successful in managing a wide range of emergencies, from wildfires to terrorism.

ICS is an "all-risk" system that has been endorsed or adopted by many emergency response organizations including:

- The National Fire Academy (NFA).
- International Association of Chiefs of Police (IACP).
- National Fire Protection Association (NFPA).

By using management best practices, ICS helps to ensure the:

- Safety of responders and others.
- Achievement of tactical objectives.
- Efficient use of resources.

In an emergency, responders may be not be working for their day-to-day supervisors, or may be working in different locations. They need to function as part of a larger system. ICS provides a standardized structure that can pull the many parts of the on-scene response together.

An ICS can be made up of many different players, such as fire, police, medical, community and State officials, and The American Red Cross.

The EOP and the Incident Command System (Continued)

The Direction and Control Annex of most EOPs describes the interface between ICS and the EOP.

The Incident Command System:

- Is a standardized management tool for meeting the demands of small or large emergency or nonemergency situations.

- Represents "best practices" and has become the standard for emergency management across the country.

- May be used for planned events, natural disasters, and acts of terrorism.

- Is a key feature of the National Incident Management System (NIMS).

As stated in NIMS, "The ICS is a management system designed to enable effective and efficient domestic incident management by integrating a combination of facilities, equipment, personnel, procedures, and communications operating within a common organizational structure, designed to enable effective and efficient domestic incident management. A basic premise of ICS is that it is widely applicable. It is used to organize both near-term and long-term field-level operations for a broad spectrum of emergencies, from small to complex incidents, both natural and manmade. ICS is used by all levels of government—Federal, State, local, and tribal—as well as by many private-sector and nongovernmental organizations. ICS is also applicable across disciplines. It is normally structured to facilitate activities in five major functional areas: command, operations, planning, logistics, and finance and administration."

ICS is not set in stone, but can be molded to fit different situations. The point of the ICS is not to give one player complete control, but to provide one system in which all of the players work together in making decisions.

ICS Basic Features

There are 14 basic features of ICS. These are:

1. **Common terminology:** ICS uses a common terminology as its base. This allows anyone from any part of the country to communicate effectively within an ICS system. Common terms for common functions, actions, and personnel prevent confusion.

    Using common terminology helps to define:

    - Organizational functions.

    - Incident facilities.

- Resource descriptions.
- Position titles.

It is important to use plain English during an incident response because often there is more than one agency involved in an incident. Ambiguous codes and acronyms have proven to be major obstacles in communications. Often, agencies have a variety of codes and acronyms that they use routinely during normal operations. Not every ten code is the same, nor does every acronym have the same meaning. When these codes and acronyms are used on an incident, confusion is often the result.

NIMS requires that all responders use "plain English," referred to as "clear text," and within the United States, English is the standard language.

ICS Basic Features (Continued)

2. <mark>Modular organization:</mark> The ICS organizational structure can contract or expand depending on the magnitude of the incident or the operational necessity. <mark>The structure includes five functional areas</mark>:

   - Command
   - Operations
   - Planning
   - Logistics
   - Finance/Administration

   Because ICS is a modular organization, it:

   - Develops in a top-down, modular fashion
   - Is based on the size and complexity of the incident.
   - Is based on the hazard environment created by the incident.

   When needed, separate functional elements can be established, each of which may be further subdivided to enhance internal organizational management and external coordination.

   <mark>Employing a modular organization means that:</mark>

   - Incident objectives determine the organizational size.
   - Only functions/positions that are necessary will be filled.
   - Each element must have a person in charge.

ICS Basic Features (Continued)

3. <u>Management by Objectives</u>: Another key ICS feature, this simply means that:

   - ICS is managed by objectives.
   - Objectives are communicated throughout the entire ICS organization through the incident planning process.

   Incident objectives are established on the following priorities:

   First Priority: Life Safety
   Second Priority: Incident Stabilization
   Third Priority: Property Preservation

4. <u>Reliance on an Incident Action Plan (IAP)</u>: Every incident must have an IAP that:

   - Specifies the incident objectives.
   - States the activities to be completed.
   - Covers a specified timeframe, called an operational period.
   - May be oral or written—except for hazardous materials incidents, which require a written IAP.

5. <u>Chain of Command and Unity of Command:</u> Together, these principles help to clarify reporting relationships and eliminate the confusion caused by multiple, conflicting directives. Incident managers at all levels must be able to control the actions of all personnel under their supervision.

   <u>Chain of command</u> is an orderly line of authority with the ranks of the incident management organization.

   Under <u>unity of command,</u> personnel:

   - Report to only one supervisor.
   - Maintain formal communication relationships only with that supervisor.

Unit 6: Planning and Coordination

ICS Basic Features (Continued)

6. **Unified Command:** The Unified Command structure:

   - Enables all responsible agencies to manage an incident together by establishing a common set of incident objectives and strategies.
   - Allows Incident Commanders to make joint decisions by establishing a single command structure.
   - Maintains unity of command. Each employee reports to only one supervisor.

7. **Manageable Span of Control:** Supervisors must be able to adequately supervise and control their subordinates, as well as communicate with and manage all resources under their supervision.

   Span of control:

   - Pertains to the number of individuals or resources that one supervisor can manage effectively during an incident.
   - Is key to effective and efficient incident management.

   The ICS span of control for any supervisor:

   - Is between 3 and 7 subordinates.
   - Optimally does not exceed 5 subordinates.

   The ICS modular organization can be expanded or contracted to maintain an optimal span of control.

ICS Basic Features (Continued)

8. <u>Pre-designated Incident Facilities</u>: Various types of operational locations and support facilities are established in the vicinity of an incident to accomplish a variety of purposes, such as decontamination, donated goods processing, mass care, and evacuation. ICS uses pre-designated incident facilities, established by the Incident Commander based on the requirements and complexity of the incident.

   Facilities may include:

   - <u>Incident Command Post (ICP):</u> The field location at which the primary tactical-level on-scene incident command functions are performed.

   - <u>Base:</u> The location at which primary Logistics functions for an incident are coordinated and administered. There is only one base per incident.

   - <u>Staging Area(s)</u>: Location established where resources can be placed while awaiting a tactical assignment.

   - <u>Camp</u>: A geographical site, within the general incident area, separate from the Incident area, equipped and staffed to provide sleeping, food, water, and sanitary services to incident personnel.

9. <u>Resource Management</u>: Resources at an incident must be managed effectively. Maintaining an accurate and up-to-date picture of resource utilization is a critical component of incident management.

   Resource management includes processes for:

   - Categorizing resources.
   - Ordering Resources.
   - Dispatching resources.
   - Tracking resources.
   - Recovering resources.
   - Reimbursement for resources, as appropriate.

   In ICS, resources are defined as personnel, teams, equipment, supplies, and facilities.

Unit 6: Planning and Coordination

ICS Basic Features (Continued)

10. **Information and Intelligence Management:** It is important that the incident management organization establishes a process for gathering, sharing, and managing incident-related information and intelligence.

    The following are examples of information and intelligence used to manage an incident:

    - Risk assessments
    - Medical intelligence (i.e., surveillance)
    - Weather information
    - Geospatial data
    - Structural designs
    - Toxic contaminant levels
    - Utilities and public works data

11. **Integrated Communications:** It is important to develop an integrated voice and data communications system before an incident.

    Incident communications are facilitated through the:

    - Development and use of a common communications plan.
    - Interoperability of communication equipment, procedures, and systems.

    Types of resources that are available may include:

    - Radio systems and frequencies.
    - Telephone systems.
    - Computers
    - Message runners, coding, and signaling.

ICS Basic Features (Continued)

12. <mark>Transfer of Command:</mark> The process of moving responsibility for incident command from one Incident Commander to another is called transfer of command. Transfer of command must include a transfer of command briefing—which may be oral, written, or a combination of both.

    Transfer of command occurs when:

    - A more qualified person assumes command.

    - The incident situation changes over time, resulting in a legal requirement to change command.

    - There is normal turnover of personnel on extended incidents.

    - The incident response is concluded and responsibility is transferred to the home agency.

13. <mark>Accountability:</mark> Effective accountability during incident operations is essential. Individuals must abide by their agency policies and guidelines and any applicable local, State, or Federal rules and regulations.

    The following principles must be adhered to:

    - <mark>Check-in.</mark> All responders must report in to receive an assignment in accordance with the procedures established by the Incident Commander.

    - <mark>Incident Action Plan.</mark> Response operations must be coordinated as outlined in the IAP.

    - <mark>Unity of Command.</mark> Each individual will be assigned to only one supervisor.

    - <mark>Span of Control.</mark> Supervisors must be able to supervise and control their subordinates and communicate with and manage all resources under their supervision.

    - <mark>Resource Tracking.</mark> Supervisors must record and report resource status changes as they occur.

Unit 6: Planning and Coordination

ICS Basic Features (Continued)

14. **Mobilization:** It is important to manage resources to adjust to changing conditions.

    At any incident:

    - The situation must be assessed and the response planned.
    - Managing resources safely and effectively is the most important consideration.
    - Personnel and equipment should respond only when requested or when dispatched by an appropriate authority.

The EOP and Multi Agency Coordination/Emergency Operations Center

Multiagency coordination is a process that allows all levels of government and all disciplines to work together more efficiently and effectively. Multiagency coordination occurs across the different disciplines involved in incident management, across jurisdictional lines, or across levels of government.

MACS is a system . . . not simply a facility.

Multiagency coordination can and does occur on a regular basis whenever personnel from different agencies interact in such activities as preparedness, prevention, response, recovery, and mitigation. Often, cooperating agencies develop a MACS to better define how they will work together and to work together more efficiently; however, multiagency coordination can take place without established protocols. MACS may be put in motion regardless of the location, personnel titles, or organizational structure. MACS includes planning and coordinating resources and other support for planned, notice, or no-notice events. MACS defines business practices, standard operating procedures, processes, and protocols by which participating agencies will coordinate their interactions. Integral elements of MACS are dispatch procedures and protocols, the incident command structure, and the coordination and support activities taking place within an activated EOC. Fundamentally, MACS provide support, coordination, and assistance with policy-level decisions to the ICS structure managing an incident.

The two most commonly used elements of the Multiagency Coordination System are EOCs and MAC Groups.

An EOC is a central location where agency representatives can coordinate and make decisions when managing an emergency response. The EOP designates the facility that will serve as the EOC during emergencies. Specifying an EOC allows decision makers to operate in one place to coordinate and communicate with support staff.

The advantages of a single EOC location include:

- Centralized priority setting, decision-making, and resource coordination.
- Simplified long-term operation.
- Increased continuity.
- Better access to all available information.
- Easier verification of information.
- Easier identification and deployment of available resources.

The EOC should be located away from vulnerable, high-risk areas but accessible to the local officials who will use it. A convenient, secure location will:

- Provide a single, recognizable focal point for emergency or disaster management.
- Allow emergency organizations to respond as a team.
- Permit a faster response and recovery than a fragmented approach would provide.

Also, a single facility can function more efficiently because calls for assistance can be made to a single location where key officials can:

- Meet.
- Make decisions.
- Coordinate activities.

The EOP and the EOC (Continued)

The EOC/MAC does not provide on-scene management but manages the overall event through seven key functions:

The Multiagency Coordination System should be both flexible and scalable to be efficient and effective. MACS will generally perform common functions during an incident; however, not all of the system's functions will be performed during every incident, and functions may not occur in any particular order.

1. Situation Assessment
This assessment includes the collection, processing, and display of all information needed. This may take the form of consolidating situation reports, obtaining supplemental information, and preparing maps and status boards.

2. Incident Priority Determination
Establishing the priorities among ongoing incidents within the defined area of responsibility is another component of MACS. Typically, a process or procedure is established to coordinate with Area or Incident Commands to prioritize the incident demands for critical resources. Additional considerations for determining priorities include the following:
- Life-threatening situations.
- Threat to property.
- High damage potential.
- Incident complexity.
- Environmental impact.
- Economic impact.
- Other criteria established by the Multi-agency Coordination System.

3. Critical Resource Acquisition and Allocation
Designated critical resources will be acquired, if possible, from the involved agencies or jurisdictions. These agencies or jurisdictions may shift resources internally to match the incident needs because of incident priority decisions. Resources available from incidents in the process of demobilization may be shifted, for example, to higher priority incidents.
Resources may also be acquired from outside the affected area. Procedures for acquiring outside resources will vary, depending on such things as the agencies involved and written agreements.

4. Support for Relevant Incident Management Policies and Interagency Activities
A primary function of MACS is to coordinate, support, and assist with policy-level decisions and interagency activities relevant to incident management activities, policies, priorities, and strategies.

5. Coordination With Other MACS Elements

A critical part of MACS is outlining how each system element will communicate and coordinate with other system elements at the same level, the level above, and the level below. Those involved in multi-agency coordination functions following an incident may be responsible for incorporating lessons learned into their procedures, protocols, business practices, and communications strategies. These improvements may need to be coordinated with other appropriate preparedness organizations.

6. Coordination With Elected and Appointed Officials

Another primary function outlined in MACS is a process or procedure to keep elected and appointed officials at all levels of government informed. Maintaining the awareness and support of these officials, particularly those from jurisdictions within the affected area, is extremely important, as scarce resources may need to move to an agency or jurisdiction with higher priorities.

7. Coordination of Summary Information

By virtue of the situation assessment function, personnel implementing the multiagency coordination procedures may provide summary information on incidents within their area of responsibility as well as provide agency/jurisdictional contacts for media and other interested agencies.

The Multi-Agency Coordination System integrates facilities, equipment, personnel, procedures and communications into a common system with responsibility for coordinating and supporting domestic incident management activities. The functions of the system are to support incident management polities and procedures, facilitate logistical support and resource tracking, inform resource allocation decisions, coordinate incident-related information, and coordinate interagency and intergovernmental issues regarding policies, priorities, and strategies.

Unit 6: Planning and Coordination

Activity: The EOP, ICS, and EOC

The purpose of this activity is to match each feature to the EOP, ICS, or EOC. For every feature in Column A, choose the appropriate category for that feature from Column B.

The EOP, ICS, and EOC	
1. __B__ Modular organization	a. EOP
2. __C__ Single location	b. ICS
3. __A__ Responsibilities for core emergency management functions	c. EOC
4. __B__ Comprehensive resource management	
5. __A__ Roles of participating agencies	
6. __B__ Method of working together	
7. __B__ Unified command structure	

Fundamentals of Emergency Management

UNIT 6: PLANNING AND COORDINATION

Activity: The EOP, ICS, and the EOC (Continued)

Answers to the Activity

1. b
2. c
3. a
4. b
5. a
6. b
7. b

Unit 6: Planning and Coordination

 Case Study: Multiple-Agency Coordination

The following description of the response to an elevator grain explosion illustrates the complex coordination among the EOP, ICS, EOC, and individual agencies. When you have finished reading the case study, complete the questions on the following page.

A grain elevator exploded in DeBruce, Kansas, a largely rural area, in 1998. There were 7 fatalities, 10 people injured, and 30 who had to be evacuated from the scene.

Sedgwick County and the Kansas State Governor declared a State of Emergency, and the disaster received a Presidential Declaration of Emergency. An Urban Search and Rescue Task Force was dispatched to the site.

The following local, State, and Federal personnel were involved in the response:

- 4 appointed officials
- 22 communications personnel
- 10 elected officials
- 73 emergency management personnel
- 250 fire/rescue personnel
- 12 health and medical specialists
- 9 human services personnel
- 62 law enforcement officers
- 28 military personnel
- 4 public relations officers
- 33 public works personnel
- 10 utility personnel
- 20 volunteers

The scene remained open for response operations over a period of 1 month and 2 days.

Unit 6: Planning and Coordination

Case Study: Multiple-Agency Coordination (Continued)

How did all of these responders know who was in charge at any given time?

What source did communications personnel consult for correct technical protocols and procedures?

Where did elected officials and agency heads meet to make policy decisions?

UNIT 6: PLANNING AND COORDINATION

Case Study: Multiple-Agency Coordination (Continued)

Answers to Case Study

How did all of these responders know who was in charge at any given time?

The ICS provides for unity of command. The Direction and Control Annex of the EOP specifies how the ICS is implemented.

What source did communications personnel consult for correct technical protocols and procedures?

Their agencies' SOPs.

Where did elected officials and agency heads meet to make policy decisions?

In the EOC.

Augmenting Local Resources

As mentioned earlier, resources for an integrated emergency management system include both personnel and equipment.

Response to an incident, such as the grain elevator explosion, called for far more personnel and equipment than local agencies could provide. Most ongoing emergency management programs anticipate possible resource shortages, and seek ways in which to augment them.

Some strategies for obtaining additional resources include:

- Pre-emergency purchase and storage of items that are known to be needed during an emergency, but are not currently in inventory (e.g., chain saws and other tools, plywood, plastic sheeting, drinking water).

- Standby contracts, which allow the community to purchase or lease needed items (ranging from plastic sheeting to heavy equipment) at a price equal to the price in effect on the day before the emergency.

- Mutual aid and assistance agreements with other neighboring communities, local, and State governments, in which resources from those groups are transported to the community experiencing the emergency, then returned when they are no longer needed.

- Solicited donations.

- Resource typing to quickly and effectively identify, locate, request, order, and track outside resources.

The Resource Management Annex to the EOP directs how resource needs will be met during response to an emergency.

Augmenting Local Resources (Continued)

As resource shortages occur, the resource management staff at the EOC receives reports of any needs that cannot be met with an agency's resources. The Resource Management staff gathers essential information before trying to fulfill the needs. This information includes:

- What is needed.
- How it will be used. (Information should be as specific as possible because a different item might work as well or better and be readily available).
- How much is needed.
- Who needs it.
- Where it is needed.
- When it is needed.

It is important for EOC staff to set priorities when meeting needs. After a resource need is identified, it should be logged, passed on to those responsible for obtaining and committing resources, and then tracked.

Pre-emergency Purchase

Some communities make the decision to purchase and store items that are known to be needed in an emergency. The main advantage of purchasing items in advance of an emergency is that, when an emergency occurs, the items may not be available or may be available at a much higher price.

There are two key disadvantages to pre-emergency purchase, however:

- Purchase ties up funds for items that may not be used within a known timeframe. Emergency Managers and community leaders will have to determine whether the benefit of pre-emergency purchase outweighs the cost—especially for items that require controlled storage or that deteriorate over time.

- Purchase is not the only cost involved with the resources. The resources must be inventoried, stored, and maintained, adding an additional cost burden to the community.

After considering the costs versus the benefits of pre-emergency purchase, many communities opt for standby contracts as a more cost-effective alternative.

Standby Contracts

Standby contracts allow communities to purchase or lease items needed for an emergency response at the price in effect on the day before the emergency occurred. These contracts offer several advantages over pre-emergency purchase:

- They ensure that the resources required will be available within a specified timeframe and at an established price.
- They eliminate the need for inventory, storage, and maintenance that accompanies pre-emergency purchase.

A potential disadvantage exists with standby contracts if, in the aftermath of an emergency, the local infrastructure is so disrupted that accessing and distributing the contracted materials becomes a logistical nightmare. Additionally, in a widespread emergency, suppliers may be overextended and unable to deliver the supplies as agreed to in the contract.

Mutual Aid and Assistance Agreements

Most emergencies and disasters do not receive Federal disaster declarations. Developing mutual aid and assistance agreements with adjacent communities can be important to ensuring that adequate resources are available to address an emergency.

In any emergency or disaster, mutual aid partners may be able to provide:

- Emergency personnel.
- Equipment, such as bulldozers or dump trucks.
- Communications capability.
- Overall management strategy and program management.
- Sandbags.
- Facilities, such as warehouses or temporary shelters.

Mutual Aid Agreements (Continued)

Mutual aid and assistance agreements usually are documented in the Resource Management Annex to the EOP. Although designed for emergencies, emergency management partnerships should operate outside of a disaster setting as well. Training, preparedness, and mitigation efforts also can be shared and enhanced through partnerships

Donations

Solicited Donations

During some types of emergencies, it may be possible to solicit donations of needed supplies from suppliers or directly from the public. Typically, solicited donations involve items such as four-wheel drive transportation following a blizzard or boats following a flood. Soliciting other types of donations, such as emergency supplies, can cause major problems, however.

Unsolicited Donations

After-action reports are full of issues relating to unsolicited donations that arrive at a disaster site in trailers that are filled with unsorted, unneeded goods. Even in communities that have an established mechanism for dealing with donated goods, unsolicited donations create huge logistical problems. Most communities, therefore, prefer to request cash instead of goods.

Maintaining an Effective EOP

Regardless of how good an EOP is, it will not work if it is not communicated. At a minimum, the plan must be communicated to:

- Local, State, and Federal officials who need to coordinate the plan with their EOPs.

- Response personnel both inside and outside of the community who share responsibility for implementing the plan, reducing damage, and saving lives.

- The local community, which has expectations concerning the government's role in an emergency and, collectively, is critical to the plan's success.

Maintaining an Effective EOP (Continued)

The best way to communicate the plan to response agencies that are responsible for implementing it is through training and exercising.

Training is critical to response personnel so that they know:

- What they are supposed to do.
- When they are to do it.
- How they are to do it, including:
    - Procedures for accomplishing their task or mission.
    - Coordinating their efforts with personnel within and outside of the agency.
    - Communicating their needs and status.

Training can include a wide range of activities from classroom training; to on-the-job training; to the use of checklists, worksheets, and job aids. The type and duration of the training selected depends on the:

- Frequency with which the task is performed. (Tasks that responders perform often will require less training than tasks that they perform only during an emergency situation.)
- The complexity of the task.

Exercises are critical to a plan's success and a successful response because they:

- Test and evaluate plans, policies, and procedures.
- Identify planning weaknesses.
- Identify resource gaps.
- Improve interagency coordination and communication.

Maintaining an Effective EOP (Continued)

Exercises will show whether what appears to work on paper actually *does* work in practice. Exercising will help to:

- Clarify the roles and responsibilities of all who play any part in the response.

- Improve individual performance by providing an opportunity for responders and others to practice their assigned duties.

- Gain public recognition that the local government has taken steps to protect their safety—and gain the support of public officials who will support the response effort during an emergency.

There are several types of exercises graduating in realism, complexity, and stress levels.

Interfacing with Other Plans

Many cities and counties employ professional planners to develop and maintain comprehensive plans for their areas. A comprehensive plan includes a study of the traffic and transportation characteristics, population, economy and sociology, and the physical features of the community.

Emergency plans and comprehensive plans have obvious overlaps. While there is a mitigation component in local emergency management programs, comprehensive plans also address mitigation. For example, the land use element of the comprehensive plan may call for acquisition of park land where property is vulnerable to residential development within a floodplain. The community's land use plans may specify the location of future growth and development, as well as the stated goals and policies of the community. Regulations serve as tools for implementing plans and managing development.

Land use planning considers the impact of changes in land use and development that also change the hazard profile of the community. Potable water, sewer service, roads, storm water runoff patterns, and water quality all may be affected by development. Storms and erosion also alter a community's hazard areas.

Interfacing with Other Plans (Continued)

Communities can incorporate emergency management policies and goals into the local planning process through enactment of regulations and policies that support structural and nonstructural mitigation measures. A close working relationship between the emergency planning and comprehensive planning functions strengthens both programs.

One critical area of interface between the EOP and other plans is in the hazardous materials arena.

- Hazardous materials are accompanied by their own planning requirements, as established by the Department of Transportation (DOT) in 49 Code of Federal Regulations (CFR), and the Occupational Safety and Health Administration (OSHA) in 29 CFR.

- The Environmental Protection Agency (EPA) places additional requirements for hazardous waste under 40 CFR.

These regulations cover a range of requirements, including:

- Transport and storage of hazardous materials, including placard requirements (49 CFR).

- Hazard communication under SARA Title III (29 CFR).

- Clean up and disposal of hazardous materials (49 CFR) and hazardous waste (40 CFR).

Additionally, these regulations require that all incidents involving hazardous materials or hazardous waste be managed using ICS.

If you are unfamiliar with planning requirements related to hazardous materials or hazardous waste, consult with your HazMat officers, Local Emergency Planning Committee (LEPC), or State emergency management agency to ensure that your local EOP also meets the requirements for hazardous materials and hazardous waste response.

Unit 6: Planning and Coordination

Knowledge Check

Carefully read each question and all of the possible answers before selecting the most appropriate response for each test item. Circle the letter corresponding to the answer you have chosen.

1. Standby contracts establish the price as that which is/was in effect:

   a. On the day before the emergency event.
   b. On the day of the emergency event.
   c. On the day after an emergency event.

2. The Incident Command System provides a model for command, control, and coordination of a response operation.

   a. True
   b. False

3. The EOP provides _____ during emergencies.

   a. A method of working together
   b. Centralized direction and control
   c. A source of overall authority, roles, and functions
   d. Communications capabilities
   e. Recommended mitigation measures

4. Which of the following is not a feature of the Incident Command System?

   a. Central location
   b. Incident Action Plans
   c. Manageable span of control
   d. Common terminology
   e. Modular organization

5. Mutual aid and assistance agreements may provide _____ when invoked.

   a. Unity of command
   b. Emergency public information
   c. Policy decisions
   d. Training
   e. Equipment such as bulldozers or dump trucks

Fundamentals of Emergency Management

UNIT 6: PLANNING AND COORDINATION

Knowledge Check (Continued)

Answers to Knowledge Check

1. a
2. a
3. c
4. a
5. e

Unit 7: Functions of an Emergency Management Program

UNIT 7: FUNCTIONS OF AN EMERGENCY MANAGEMENT PROGRAM

Introduction and Unit Overview

This unit examines the functions of an emergency management program. After completing this unit, you should be able to:

- Describe the emergency management core functions that are performed during emergencies.
- Describe the emergency management program functions that continue on a day-to-day basis.
- Distinguish between core functions and program functions.
- Discuss the role of local laws in establishing emergency management authorities and responsibilities.

Introduction to Emergency Management Functions

Emergency management is most visible during emergencies, but successful response operations require continual operations between incidents.

There are two ways to categorize emergency management activities:

1. Emergency management core functions that are performed during emergencies.
2. Emergency management program functions that continue on a day-to-day basis.

In this unit you will learn about the emergency core functions and the essential functions of an ongoing emergency management program.

UNIT 7: FUNCTIONS OF AN EMERGENCY MANAGEMENT PROGRAM

Basis in Local Law

Specific areas of authority and responsibilities for emergency management should be clearly stated in local ordinances. Local ordinances should spell out who has responsibility for:

- Emergency management operations in normal, day-to-day activities.
- Policy decisions affecting long-term emergency management.
- Final authority in actual disaster situations.

These ordinances should also stipulate when emergency authorities begin during an emergency and end following the response.

Local laws:

- Provide for a specific line of succession for elected officials and require that departments of government establish lines of succession. This ensures continuity of government and leadership in an emergency.
- Define and delineate responsibilities, scopes of authority, and standards for the position of emergency program manager for an all-hazards integrated local emergency plan, and for mutual support.

Emergency Management Core Functions

The Eight Core Functions are:

1. Direction and Control.
2. Communications.
3. Warning.
4. Emergency Public Information.
5. Evacuation.
6. Mass Care.

Emergency Management Core Functions (Continued)

7. Health and Medical.

8. Resource Management.

Assigning work based on these core functions helps to ensure that continuity with the EOP is maintained during a response.

Note that some State and local EOPs use different terms for the eight core functions. If that is the case in your State, review the functions in your EOP to see where each of the core functions from SLG-101 are covered.

Emergency Management Program Functions

Most emergency management reports and surveys are organized according to a standard set of emergency management functions. The functions serve as a means to divide day-to-day program activities into categories. The Emergency Manager uses these functions to organize and direct the emergency management program.

The standard emergency management program functions used by most local governments are shown in the following table.

Emergency Management Program Functions (Continued)

Emergency Management Program Functions	
Function	Description
1. Laws and Authorities	A legal basis for the establishment of the emergency management organization, the implementation of an emergency management program, and continuity of government exists in local law/ordinance and is consistent with State statutes concerning emergency management.
2. Hazard Identification and Risk Assessment	The jurisdiction has a method for identifying and evaluating natural and technological hazards within its jurisdiction.
3. Hazard Mitigation	The jurisdiction has established a pre-disaster hazard mitigation program.
4. Resource Management	The local emergency management organization has the human resources required to carry out assigned day-to-day responsibilities.
5. Planning	The jurisdiction has developed a comprehensive mitigation plan and an EOP.
6. Direction and Control	EOC operating procedures are developed and tested annually.
7. Communication and Warning	Communications system capabilities are established.
8. Operations and Procedures	The jurisdiction has developed procedures for conducting needs and damage assessments, requesting disaster assistance, and conducting a range of response functions.
9. Logistics and Facilities	The primary and alternate EOCs have the capabilities to sustain emergency operations for the duration of the emergency and have developed logistics management and operations plans.
10. Training	The jurisdiction conducts an annual training needs assessment, incorporates courses from various sources, and provides/offers training to all personnel with assigned emergency management responsibilities.
11. Exercises, Evaluations, and Corrective Actions	The jurisdiction has established an emergency management exercises program, exercises the EOP on an annual basis, and incorporates an evaluation component and corrective action program.

Emergency Management Program Functions (Continued)

Emergency Management Program Functions	
Function	Description
12. Public Education and Information	An emergency preparedness public education program is established, procedures are established for disseminating and managing emergency public information in a disaster, and procedures are developed for establishing and operating a Joint Information Center (JIC).
13. Finance and Administration	The jurisdiction has established an administrative system for day-to-day operations.

UNIT 7: FUNCTIONS OF AN EMERGENCY MANAGEMENT PROGRAM

Case Study: Train Derailment Review

The train derailment and anhydrous ammonia release incident that was described earlier in this course is recapped below. You will use the facts to identify how each of the emergency management program functions could be applied to this event.

The freight train derailed at approximately 3:00 AM, releasing anhydrous ammonia, a toxic gas.

The Chairman of the County Board of Supervisors declared a State of Emergency, activating the local EOP. The local EOP includes the county and all incorporated towns and cities within the county. There are mutual aid agreements with surrounding counties.

The EOC opened, and policymakers gathered to direct the response. Because warning systems that rely on radio or television transmission would fail to alert most people, warning sirens were used.

Any evacuation attempt would have exposed residents directly to the hazard, so they were initially advised to shelter in their homes. Eventually 21 homes were evacuated. One resident perished while attempting to leave the area.

Approximately one third of a nearby city was also affected, but residents were not able to evacuate. All those affected were advised to shelter in place.

There were some delays activating responders, who could not enter the accident vicinity without proper gear.

One responder was trapped after he drove into a ditch trying to leave the scene because his vehicle windshield was coated with frozen gas from the toxic cloud. The responder was rescued some time later.

Hazardous Materials appendixes in the EOP listed sources for protective gear, procedures for working with hazardous materials, protective actions that could be taken, and information that should be given to the public. The Emergency Management Agency had provided training on hazardous materials to local responders, and a disaster simulation involving a hazardous chemical release is scheduled.

The Emergency Management Agency followed appendix procedures by activating warning sirens and broadcasting instructions to "shelter in place" by closing all windows and turning off furnaces to avoid bringing outside air into buildings.

Media attention was intense. Survivors needed public information on treating exposure symptoms, cleaning homes, and dealing with exposed pets and livestock. Many horses, being especially sensitive to airborne contaminants, died.

Case Study: Train Derailment Review (Continued)

Transportation of hazardous materials is Federally regulated and imposes requirements on carriers. The railroad contracted with a HazMat team to clean up the site.

The railroad also contacted those with hospital bills and assumed responsibility for payment. The railroad had survivors sign releases of liability for additional damages. The State eventually forced the railroad to cease requiring releases.

The State Health Department has been monitoring air and water quality in the area. Some contaminated dirt has been removed.

UNIT 7: FUNCTIONS OF AN EMERGENCY MANAGEMENT PROGRAM

Activity: Emergency Management Functions in Action

Jot down one way that each emergency management function actually applied or could apply to the train derailment and hazardous materials release described earlier. For example, the Laws and Authorities function established the legal basis for the emergency response by the involved jurisdictions.

Emergency Management Function	Application to the Derailment/Anhydrous Ammonia Release
1. Laws and Authorities	
2. Hazard Identification and Risk Assessment	
3. Hazard Mitigation	
4. Resource Management	
5. Planning	
6. Direction and Control	
7. Communication and Warning	
8. Operations and Procedures	
9. Logistics and Facilities	
10. Training	
11. Exercises, Evaluations, and Corrective Actions	
12. Public Education and Information	
13. Finance and Administration	

Fundamentals of Emergency Management

UNIT 7: FUNCTIONS OF AN EMERGENCY MANAGEMENT PROGRAM

Activity: Emergency Management Functions in Action (Continued)

Here are some sample answers. There are many possible applications for each of the functions.

Emergency Management Function	Application to the Derailment/Anhydrous Ammonia Release
1. Laws and Authorities	The transportation of hazardous materials is Federally regulated, so Federal regulations affect the local response.
2. Hazard Identification and Risk Assessment	Transportation of hazardous materials close to population centers causes risk of releases. The EOP should have an appendix dealing with hazardous materials that address this risk.
3. Hazard Mitigation	Zoning changes and better management by the railroad are possible measures.
4. Resource Management	There were delays in getting protective gear for responders.
5. Planning	A single plan that tied together county and city responders avoided conflicts due to competing emergency plans.
6. Direction and Control	Policymakers gathered in the EOC to establish the overall direction of the response.
7. Communication and Warning	Warning sirens were used.
8. Operations and Procedures	Each department listed in the plan was notified, and alerted its employees and volunteers.
9. Logistics and Facilities	The EOC was activated.
10. Training	Local responders had received training in dealing with hazardous materials releases.
11. Exercises, Evaluations, and Corrective Actions	A simulation that includes a hazardous materials release is scheduled.
12. Public Education and Information	Residents were advised to shelter in place, and also received information on treating exposure symptoms, cleaning homes, and dealing with exposed pets and livestock
13. Finance and Administration	The Emergency Management Agency maintained records of expenses for possible compensation by the railroad.

Fundamentals of Emergency Management

UNIT 7: FUNCTIONS OF AN EMERGENCY MANAGEMENT PROGRAM

Activity: Comparing Functions

Check one or both columns to show which functions are part of the eight core functions of emergency management, and which are functions of an emergency management program.

Function	Core Function	Program Function
Operations and Procedures		✓
Logistics and Facilities		✓
Mass Care	✓	
Direction and Control	✓	✓
Training		✓
Communications	✓	✓
Exercises, Evaluations, and Corrective Actions		✓
Public Education and Information		✓
Evacuation or Sheltering In-Place	✓	
Health and Medical	✓	
Finance and Administration		✓
Warning	✓	
Laws and Authorities		✓
Hazard Identification and Risk Assessment		✓
Hazard Mitigation		✓
Emergency Public Information	✓	
Resource Management	✓	
Planning		✓
Communication and Warning	✓	✓

UNIT 7: FUNCTIONS OF AN EMERGENCY MANAGEMENT PROGRAM

Activity: Comparing Functions (Continued)

Function	Core Function	Program Function
Operations and Procedures		✓
Logistics and Facilities		✓
Mass Care	✓	
Direction and Control	✓	✓
Training		✓
Communications	✓	
Exercises, Evaluations, and Corrective Actions		✓
Public Education and Information		✓
Evacuation or In-Place Sheltering	✓	
Health and Medical	✓	
Finance and Administration		✓
Warning	✓	
Laws and Authorities		✓
Hazard Identification and Risk Assessment		✓
Hazard Mitigation		✓
Emergency Public Information	✓	
Resource Management	✓	✓
Planning		✓
Communication and Warning		✓

Fundamentals of Emergency Management

UNIT 7: FUNCTIONS OF AN EMERGENCY MANAGEMENT PROGRAM

Knowledge Check

Carefully read each question and all of the possible answers before selecting the most appropriate response for each test item. Circle the letter corresponding to the answer you have chosen.

1. Which statement is true?

   a. Emergency program functions and emergency management core functions are totally separate.
   b. Emergency program functions can be emergency management functions at the discretion of the emergency manager.
   c. All emergency program functions also are emergency management core functions.
   d. Some emergency program functions also are emergency management core functions.

2. The 13 emergency management program functions are the categories reported in Capabilities Assessment for Readiness surveys.

   a. True
   b. False

3. As an emergency management core function, _____ is defined as a process to quickly procure, distribute, and utilize personnel and materials needed in an emergency.

   a. Direction and Control
   b. Logistics and Facilities
   c. Resource Management
   d. Communications

4. As an emergency management program function, Resource Management is described as _____.

   a. The human resources required to carry out assigned day-to-day responsibilities
   b. An administrative system for day-to-day operation
   c. A process to quickly procure, distribute, and utilize personnel and materials needed in an emergency
   d. Meeting the needs of the population despite disruption of commerce and infrastructure

5. In the train derailment case study, what function served to provide overall leadership to the response effort?

   a. Resource Management
   b. Laws and Authorities
   c. Direction and Control
   d. Operations and Procedures

Fundamentals of Emergency Management

UNIT 7: FUNCTIONS OF AN EMERGENCY MANAGEMENT PROGRAM

Knowledge Check (Continued)

Answers to Knowledge Check

1. d
2. a
3. c
4. a
5. c

Unit 8: Applying Emergency Management Principles

UNIT 8: APPLYING EMERGENCY MANAGEMENT PRINCIPLES

Introduction and Unit Overview

In the previous unit, you learned about local, State, and Federal participants in emergency management programs. In this unit, you will learn how to apply the principles you have learned up until now.

Applying the Integrated Emergency Management System

The force behind integrated emergency management must be the desire to build partnerships for safer communities.

The approach to an integrated emergency management system includes:

- All hazards.
- All resources.
- All jurisdictions.
- All emergency management phases.

An integrated approach to emergency management is based on solid general management principles, with the common theme of protecting life and property.

To achieve a truly integrated system, local, State, and Federal governments, as well as private sector agencies and individuals and families must share responsibility for applying resources effectively at every stage and phase of emergency management. While each group, unit, and individual in the system has its own role and function, the ultimate responsibility is shared among all. The result of their joint effort is a team product that reflects the insights, experiences and skills of the entire team.

UNIT 8: APPLYING EMERGENCY MANAGEMENT PRINCIPLES

Activity: Interdependence Within the Emergency Management Team

This activity presents a structured format with which to explore relationships among emergency personnel in various programs and functional areas. Because a central goal of this course is to promote interrelationships, the effort you devote to exploring interdependence is especially valuable.

1. To begin, choose a role to play (not your own) from the following list of roles.

   - Local Fire Chief
   - Local Executive Officer (Chief Elected Official)
   - Chief of Emergency Medicine at the Local Hospital
   - State Director of Emergency Services
   - Superintendent of Schools
   - Local Public Information Officer
   - Red Cross Disaster Director
   - Hazardous Chemicals Safety Officer at Local Plant
   - Vice-President of Local Utility Company
   - Vice President for Operations of Major Regional Rail Freight Carrier
   - Local Police Chief

2. Fill out the Interdependence Worksheets on the following pages. The worksheets will ask you to consider the following factors of the role you chose:

   - Emergency protection responsibilities
   - Possible contacts
   - Resource and information needs
   - Results or accomplishments of specified interactions
   - Possible effects

Activity: Interdependence Within the Emergency Management Team (Continued)

Proceed through your worksheets in order, completing each question before moving on to the next.

Your answers may be general. For example, if you know a *function* for a person or organization, but not the correct *title*, a description of the type is sufficient. Also, if you are unsure about a contact but think it would be valuable, include it. The value of this activity is in thinking as expansively as possible.

You may not have the detailed knowledge of this role that you have of your own, but you do not need in-depth knowledge for this activity—just empathy, imagination, and appreciation for the general role of that person in all-hazards management.

Use what you know about emergency management to develop answers.

Unit 8: Applying Emergency Management Principles

Activity: Interdependence Within the Emergency Management Team (Continued)

Interdependence Worksheets

Your Role:

In what sector do you operate in this position (Federal, State, local, or nongovernment)?

Briefly describe your role (in this position) in relation to:

- Prevention.

- Preparedness.

- Response.

- Recovery.

- Mitigation

Activity: Interdependence Within the Emergency Management Team (Continued)

Name ten contact points (by position or role) in various functional areas with whom you should interact. Consider contacts in all three levels of government as well as with voluntary organizations and business/industry. To generate this list, review your role statements and consider what you will need from others to accomplish that role.

Position or role of contact	In what sector is this contact?

Activity: Interdependence Within the Emergency Management Team (Continued)

Select five of the contacts you listed on the previous page. For each, name one activity you will undertake involving this contact. Then briefly describe information or resources you will need from that person. Consider needs in all four emergency management phases (preparedness, response, recovery, mitigation).

Contact	Result or Activity	Phase	Your Needs
_____	_____	_____	_____
	_____	_____	_____
	_____	_____	_____
_____	_____	_____	_____
	_____	_____	_____
	_____	_____	_____
_____	_____	_____	_____
	_____	_____	_____
	_____	_____	_____
_____	_____	_____	_____
	_____	_____	_____
	_____	_____	_____
_____	_____	_____	_____
	_____	_____	_____
	_____	_____	_____

UNIT 8: APPLYING EMERGENCY MANAGEMENT PRINCIPLES

Activity: Problem Solving In Crisis-Prone County

When you completed the Interdependence Worksheets on the preceding pages, you practiced thinking about an integrated emergency management system as a network of relationships. In this activity, you will apply your knowledge of the fundamental principles and advantages and of an integrated emergency management system. The purpose of this activity is to practice taking an integrated emergency management approach to a specific problem.

You meet a young woman at a business meeting. She is a new employee in Crisis-Prone, a county in another part of your State. She has a background in management and administration and is anxious to prove herself, yet she also is mindful of politics and diplomacy.

Six weeks later, you receive a letter from her asking for your assistance in an emergency management project.

Review this letter closely and prepare a written outline of an appropriate response. The questions that follow the letter will help you prepare the outline.

Her letter begins on the next page, and a structured guideline for outlining your response follows it. Be sure to cover, at a minimum, all issues identified in the guideline.

Unit 8: Applying Emergency Management Principles

Activity: Problem Solving In Crisis-Prone County (Continued)

CRISIS-PRONE COUNTY

Division of Administration
Office of the County Executive

Dear _____:

Since our last conversation, Crisis-Prone County has experienced a near catastrophe. As a result, the county commissioners have decided that it is time to review the county's ability to respond to emergencies and to ascertain whether there is a need for a program beyond what is provided by the county.

The task of developing a briefing for the commissioners has been assigned to me. This is an opportunity to give them their first introduction to integrated emergency management, but, as a junior analyst in the organization, I will need to rely on the expertise of my more knowledgeable colleagues to develop an outline that sufficiently explains an integrated approach to emergency management.

The event that brought this need to the attention of the commissioners demonstrates both the nature and the urgency of our situation. As you know, our county has a population of 650,000 people and is situated primarily in urban and suburban communities. A variety of transportation systems crisscross the county, presenting a considerable resource and challenge. Substantial rail traffic passes through the county on a daily basis—including commercial freight traffic that runs adjacent to the AMTRAK passenger lines.

A week ago, a freight train carrying an explosive material derailed at a major rail crossing/intersection, just as a passenger train was passing on an adjacent track. Six of the 30 cars of the freight train derailed and were precariously perched—in such a way that any measurable jostling could have caused one or all of them to fall completely off the track, and perhaps even rupture and explode. At this time, the cause of the derailment is still under investigation.

The passengers from the AMTRAK train were evacuated (after some confusion) to a nearby auditorium, where they waited for several hours before AMTRAK could make alternative plans for them.

Meanwhile, five police officers and other public officials converged on the scene. Soon they were joined by officials from the freight and passenger lines. The group of "experts" grew substantially as time passed, and a heated debate erupted as to what steps should be taken, by whom, and when. Issues of authority and liability were raised and discussed—but never resolved. The entire discussion was observed and recorded by several reporters from newspapers, radio, and television. When the discussion concluded, it had been decided that the police chief and fire chief jointly would oversee operations to ensure the safety of life and property in the area surrounding the derailment.

Activity: Problem Solving In Crisis-Prone County (Continued)

In summary, the problem was resolved without loss of life or serious damage to property. However, media coverage of the event caused a public outcry the likes of which the county has not seen for years. Public scrutiny and demands for accountability have increased markedly. The commissioners are outraged at methods demonstrated during the event and have demanded a full inquiry and explanation of the entire episode.

So, we come to my task—preparing the initial outline of the briefing for the commissioners. While I certainly do not have any direct or significant influence on the final outcome, I am convinced that good work on this project will contribute to a more constructive approach. Any advice you can provide will be helpful. While this is all new to me, I recognize the crucial nature of integrated emergency management and want very much to cover at least the fundamentals in my submission.

I look forward to your response.

                              Sincerely Yours,

                              Jane Novice
                              Junior Management Analyst
                              Crisis-Prone County

Unit 8: Applying Emergency Management Principles

Activity: Problem Solving In Crisis-Prone County (Continued)

1. What is the central problem presented in Jane Novice's letter?

2. What alternative solutions/responses are appropriate for Novice to consider?

3. Of the alternatives listed, which one do you prefer and recommend?

Unit 8: Applying Emergency Management Principles

Activity: Problem Solving In Crisis-Prone County (Continued)

4. Assuming that Novice accepts your recommendation, what strategies do you suggest for implementing it?

5. Assist Novice further by briefly outlining how your recommendations and strategies for implementation might affect a situation such as the train derailment she described.

Unit 8: Applying Emergency Management Principles

Summary and Transition

In this unit, you practiced applying emergency management principles of coordination and interdependence and explained the fundamental features of an Integrated Emergency Management System. Unit 9 will provide a summary of what you have learned in this course.

UNIT 8: APPLYING EMERGENCY MANAGEMENT PRINCIPLES

Knowledge Check

Carefully read each question and all of the possible answers before selecting the most appropriate response for each test item. Circle the letter corresponding to the answer you have chosen.

1. The Red Cross Disaster Director works in the _____ sector.

   a. Federal government
   b. State government
   c. local government
   d. nongovernment

2. The Superintendent of Schools works in the _____ sector.

   a. Federal government
   b. State government
   c. local government
   d. nongovernment

3. The central problem in Crisis-Prone County, as revealed by Jane Novice's letter, was:

   a. Media attention that embarrassed local officials.
   b. Lack of a coordinated plan for dealing with local emergencies.
   c. Disagreements between police and public officials on resolving the derailment situation.
   d. Lack of plans for dealing with evacuated AMTRAK passengers.
   e. Failure to recognize the hazard posed by rail line locations.

4. Disagreements among police officers and public officials in Crisis-Prone County could have been avoided by:

   a. Reaching agreement on direction and control of local emergencies.
   b. Appointing the Board of Commissioners to oversee emergency operations.
   c. Assuring that a Public Information Officer is always present to advise officials.
   d. Conducting training courses on conflict resolution.

5. The best outcome of an official inquiry into the derailment incident would be _____.

   a. Banning rail traffic through populated areas
   b. Identifying officials who failed to properly perform their duties
   c. Payment of damages by the railroad carrying the explosive material
   d. Determination to develop an integrated emergency management system
   e. Banning of transport of hazardous materials through populated areas

UNIT 8: APPLYING EMERGENCY MANAGEMENT PRINCIPLES

Knowledge Check (Continued)

1. d
2. c
3. b
4. a
5. d

Unit 9: Course Summary and Final Exam

Introduction and Unit Overview

This unit will briefly summarize what you have learned in the *Fundamentals of Emergency Management* Course.

At the end of this unit, you should be able to:

- Discuss the main points of the course.

- Resolve any questions that you may have about any of the materials.

When you are finished with this unit, be sure to take the Final Exam available at http://training.fema.gov/EMIWeb/IS/is230a.asp

Integrated Emergency Management System

When an emergency or disaster occurs:

- Agencies from different jurisdictions and government levels need to work together. Major emergencies and disasters ignore city, county, and State boundaries.

- Rapid decision-making is required.

Without planning and coordination, emergency operations can suffer from serious misdirection and mistakes.

An integrated emergency management system provides a conceptual framework for organizing and managing emergency protection efforts. This framework prescribes when and how local officials and agencies will work together to deal with a full range of emergencies, from natural disasters to terrorism.

Integrated Emergency Management System (Continued)

Each level of government participates in and contributes to emergency management.

- Local government has direct responsibility for the safety of its people, knowledge of the situation and personnel, and proximity to both the event and resources. Emergency Support Services are the departments of local government that are capable of responding to emergencies 24 hours a day. They include law enforcement, fire/rescue, and public works. They may also be referred to as emergency response personnel or first responders.

- State government has legal authorities for emergency response and recovery and serves as the point of contact between local and Federal governments.

- Federal government has legal authorities; fiscal resources; research capabilities, technical information and services, and specialized personnel to assist local and State agencies in responding to and recovering from emergencies or disasters.

Incident Management Actions

- Pre-incident activities, such as information sharing, threat identification, planning, and readiness exercises.

- Incident activities that include lifesaving missions and critical infrastructure support protections.

- Post-incident activities that help people and communities recover and rebuild for a safer future.

Incident Management Actions (Continued)

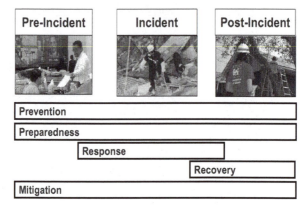

Emergency management activities, include:

Prevention: Actions taken to avoid an incident or to intervene to stop an incident from occurring, actions taken to protect lives and property, and applying intelligence and other information to a range of activities that may include countermeasures.

Preparedness: The range of deliberate, critical tasks and activities necessary to build, sustain, and improve the operational capability to prevent, protect against, respond to, and recover from domestic incidents. Preparedness is a continuous process involving efforts at all levels of government and between government and private-sector and nongovernmental organizations to identify threats, determine vulnerabilities, and identify required resources.

Response: The activities that address the short-term, direct effects of an incident. Response also includes the execution of EOPs and of incident mitigation activities designed to limit the loss of life, personal injury, property damage, and unfavorable outcomes.

Recovery: The development, coordination, and execution of service- and site-restoration plans for impacted communities and the reconstitution of government operations and services through individual, private-sector, nongovernmental, and public assistance programs.

Mitigation: Activities that are designed to reduce or eliminate risks to persons or property, or lessen the actual or potential effects or consequences of an incident.

The Plan as Program Centerpiece

Between emergencies and disasters, emergency managers can focus on mitigation and preparedness measures. Hazard analysis is a crucial first step.

Hazard analysis determines:

- What can occur.
- How often it is likely to occur.
- The devastation it is likely to cause.
- How likely it is to affect the community.
- How vulnerable the community is to the hazard.

The first step is to develop a list of hazards that may occur in the community.

Next, hazard profiles should address each hazard's:

- Duration.
- Seasonal pattern.
- Speed of onset.

The availability of Warnings also will play a crucial role in a hazard profile.

Hazard-specific information is combined with a profile of your community to determine the community's vulnerability—or risk of damage—from the hazard.

After information about the community is gathered, emergency managers use it to develop the community's hazard analysis. After a hazard and community profile has been compiled, it is helpful to quantify the community's risk by merging the information so that the community can focus on the hazards that present the highest risk.

Risk is the predicted impact that a hazard would have on people, services, and specific facilities and structures in the community. A severity rating quantifies the expected impact of a specific hazard.

Planning and Coordination

An Emergency Operations Plan (EOP) is a key component of an emergency program.

- The emergency program is in continuous operation and includes nonemergency activities, such as training and exercises.

- The EOP is activated to guide response to emergencies, and is only in effect during emergencies.

When an emergency threatens or strikes, the community must be prepared to take immediate action. An EOP describes:

- What emergency response actions will occur. . .

- Under what circumstances. . .

- Using what resources. . .

- Who will be involved and by what authority.

An EOP consists of the following components:

- The Basic Plan

- Annexes

- Appendixes

- SOPs

The EOP provides overall authority, roles, and functions during emergencies.

An Emergency Operation Center (EOC) is a central location where agency representatives can coordinate and make decisions when managing an emergency response. EOC personnel do not control the on-scene response but help on-scene personnel by establishing priorities, coordinating the acquisition and assignment of resources, and acting as a liaison with other communities and the State. The EOC is a place for working together.

The Multi-Agency Coordination System integrates facilities, equipment, personnel, procedures and communications into a common system with responsibility for coordinating and supporting domestic incident management activities. The functions of the system are to support incident management polities and procedures, facilitate logistical support and resource tracking, inform resource allocation decisions, coordinate incident-related information, and coordinate interagency and intergovernmental issues regarding policies, priorities, and strategies."

Planning and Coordination (Continued)

The Incident Command System (ICS) defines the operating characteristics, interactive management components, and structure of incident management and emergency response organizations engaged throughout the life cycle of an incident. NIMS requires the use of ICS.

Conclusion of mutual aid agreements to augment local resources is an important part of developing and maintaining an emergency management program.

In any emergency or disaster, mutual aid partners may be able to provide:

- Communications capability.
- Emergency personnel.
- Overall management strategy and program management.
- Equipment such as bulldozers or dump trucks.
- Sandbags.
- Facilities such as warehouses or temporary shelters.

The best way to communicate the plan to response agencies that are responsible for implementing the plan is through training and exercising.

Training is critical to response personnel so that they know:

- What they are to do.
- When they are to do it.
- How they are to do it.

Unit 9: Course Summary and Final Exam

Functions of an Emergency Management Program

There are two ways to categorize emergency management activities:

- Emergency management core functions that are performed during emergencies.

- Emergency management program functions that continue on a day-to-day basis.

The eight emergency management core functions performed during emergencies are:

1. Direction and control.
2. Communications.
3. Warning.
4. Emergency public information.
5. Evacuation (or in-place sheltering).
6. Mass care.
7. Health and medical.
8. Resource management.

Functions of an Emergency Management Program (Continued)

Day-to-day emergency management program functions include:

- Hazard Identification and Risk Assessment.
- Hazard Mitigation.
- Resource Management.
- Planning.
- Direction and Control.
- Communication and Warning.
- Operations and Procedures.
- Logistics and Facilities.
- Training.
- Exercises, Evaluations, and Corrective Actions.
- Public Education and Information.
- Finance and Administration.

Emergency Management Program Partners

Emergency management partners include local, State, and Federal emergency managers.

The local Emergency Program Manager has the day-to-day responsibility of managing emergency programs and activities. The role entails coordinating all aspects of a jurisdiction's, preparedness, response, recovery and mitigation capabilities.

The State's role is to supplement and facilitate local efforts before, during, and after emergencies. The State must be prepared to maintain or accelerate services and to provide new services to local governments when local capabilities fall short of disaster demands.

UNIT 9: COURSE SUMMARY AND FINAL EXAM

Emergency Management Program Partners (Continued)

The State provides direct guidance and assistance to local jurisdictions through program development, and it channels Federal guidance and assistance down to the local level. In a disaster the State office helps coordinate and integrate resources and apply them to local needs. The State's role might be best described as "pivotal."

The Federal government provides legislation, Executive Orders, and regulations that influence all disaster activities. It also maintains the largest pool of fiscal resources that can be applied to emergency management. Assistance may take the form of fiscal support, technical assistance, and information about materials, personnel resources, and research. FEMA takes a lead role in national preparedness for major crises. It also plays coordinating and supportive/assistance roles for integrated emergency management in partnership with State and local emergency management entities.

Applying Emergency Management Principles

Emergency management partners at all levels need to form interdependent networks to function as a team when responding to a given emergency situation.

You practiced describing a network of relationships connected to one emergency management system participant. You also provided a rationale for developing an integrated emergency management system in Crisis-Prone County.

Fundamentals of Emergency Management

UNIT 9: COURSE SUMMARY AND FINAL EXAM

You have now completed IS 230.a course and should be ready to take the Final Exam. You will find the IS 230.a exam questions on the next page.

Online Exam
Print out the exam and then go to http://training.fema.gov/EMIWeb/IS/is230a.asp and click on Take Final Exam to complete the exam online.

Opscan Form by Mail
To submit the final exam by mail using the Opscan Answer Sheet form, go to: http://training.fema.gov/IS/ and click on the Opscan Form Request link. Follow the instructions printed on the form and mail it to the address indicated on the form.

If you have any questions about how to take the IS 230.a exam, please call 1-301-447-1200.

UNIT 9: COURSE SUMMARY AND FINAL EXAM

Fundamentals of Emergency Management
Independent Study 230.a
Final Exam

1. The overall goal of emergency management is to:
    a. Respond quickly to all emergencies and disasters.
    b. Save lives, protect property, and protect the environment.
    c. Prevent disasters from occurring.
    d. Restore communities to pre-disaster condition.

2. The role of the local Emergency Program Manager could best be described as:
    a. Coordinating resources and activities in all phases of emergency management.
    b. Always directing all operations at the scene of an emergency.
    c. Usually working directly with the State and Federal governments.
    d. Issuing press releases and making statements to the media about disasters.

3. A request for a Presidential Declaration of Disaster must be made by the:
    a. Chief elected official of the affected area.
    b. Governor of the affected State.
    c. State Emergency Program Manager of the affected State.
    d. Designated Federal Coordinating Officer.

4. An example of a Mitigation activity is:
    a. Training.
    b. On-site operations to provide emergency assistance.
    c. Building earthquake-resistant structures in earthquake zones.
    d. Debris removal.

5. Which is the guiding document used to coordinate response and recovery actions?
    a. Standard Operating Procedures
    b. Emergency Operations Plan
    c. Risk Management Plan
    d. Community Comprehensive Plan

6. A hazard is defined as:
    a. A quantified measure of risk.
    b. A severity rating.
    c. Something that is potentially dangerous or harmful.
    d. Vulnerability to a technological hazard.

7. One of the planning factors to be considered during a hazard analysis is:
    a. Federal assistance that may be available.
    b. How quickly the community can recover.
    c. The speed of onset for each hazard.
    d. What local industry can contribute to the response?

Unit 9: Course Summary and Final Exam

8. Training and exercising is an example of a _____ activity.
   a. Mitigation
   b. Preparedness
   c. Response
   d. Recovery

9. The predicted impact that a hazard would have on people, services, and specific facilities and structures in the community is called:
   a. Hazard identification.
   b. Risk.
   c. Crisis index.
   d. Sector profile

10. An agreement between two government entities for mutual support to one another in time of emergency is called:
    a. Mutual Aid and Assistance Agreement.
    b. Services Contract.
    c. Reciprocity.
    d. Resource Sharing Contract.

11. When assessing risk, the top response priority is:
    a. Essential facilities.
    b. Critical Infrastructure.
    c. Life safety.
    d. Property conservation.

12. An emergency management program will work well in practice if most emphasis and attention focus upon
    a. A comprehensive written plan.
    b. Well-established, day-to-day relationships.
    c. Reliance on State assistance.
    d. Mutual aid and assistance.

13. If a disaster response demands more resources than any local governments can supply without assistance, the next step is to request assistance from:
    a. Congressional representatives.
    b. The President.
    c. FEMA.
    d. The State emergency management agency.

14. The first step in a hazard analysis is
    a. Identifying sources of additional resources needed to respond to each hazard.
    b. Identifying all hazards that pose a risk to your community.
    c. Determining the risk that each hazard poses to your community.
    d. Prioritizing the risk of each hazard to your community.

UNIT 9: COURSE SUMMARY AND FINAL EXAM

15. Standby contracts establish the price as that which is/was in effect:
    a. The day before the emergency event.
    b. The day of the emergency event.
    c. The month before.
    d. The day after an emergency event.

16. The Red Cross Disaster Director works in what sector?
    a. Federal government
    b. State government
    c. non-government
    d. local government

17. The local emergency program manager coordinates:
    a. The National Response Framework.
    b. The National Guard.
    c. The local emergency management programs.
    d. The disaster declaration process.

18. Who is the person in charge of an incident site?:
    a. Incident Commander
    b. Emergency manager
    c. Police chief
    d. Fire chief

19. NIMS is a
    a. Foundation that ensures all responders can work together during emergencies.
    b. Process for receiving disaster funding.
    c. National plan for terrorism response.
    d. Totally new way of doing business

20. CPG 101 provides:
    a. The foundation for State and local planning in the United States.
    b. A "fill in the blank" EOP template.
    c. Specific requirements for EOP development.
    d. Is only applicable to Federal planning.

21. The local EOC's function during an emergency is to:
    a. Give tactical direction to the incident commander.
    b. Support to the incident commander.
    c. Manage the incident.
    d. Assure span of control is not exceeded.

22. The National Response Framework:
    a. Identifies response roles, responsibilities and structures.
    b. Is only applicable to Federal Departments and Agencies.
    c. Needs to be activated before it is useful.
    d. Uses non-NIMS compliant structures.

23. Congress passed the Post-Katrina Emergency Reform Act to:
    a. Revise NIMS.
    b. Improve disaster response.
    c. Reduce Federal disaster assistance.
    d. Mandate training for first responders.

24. Integrated emergency management increases capability by:
    a. Establishing partnerships and good working relationships among all organizations involved in emergencies.
    b. Reducing interagency collaboration.
    c. Focusing on a single agency's issues.
    d. Conducting discipline specific training and exercises.

25. Individuals and Families can contribute to an emergency management program by:
    a. Assuming that the government will take care of their emergency needs.
    b. Building in floodplains.
    c. Allowing the batteries in their portable radios and flashlights to die.
    d. Developing and exercising their own disaster plan.

Appendix A: Job Aid

APPENDIX A: JOB AID

Job Aid 5.1: Hazard Analysis Worksheet

Hazard: _____

Frequency of Occurrence:

☐ *Highly likely* (Near 100% probability in the next year)
☐ *Likely* (Between 10% and 100% probability in the next year, or at least one chance in the next 10 years)
☐ *Possible* (Between 1% and 10% probability in the next year, or at least one chance in the next 100 years)
☐ *Unlikely* (Less than 1% probability in the next 100 years)

Seasonal pattern?

☐ No
☐ Yes. Specify season(s) when hazard occurs: _____

Potential Impact:

☐ *Catastrophic* (Multiple deaths; shutdown of critical facilities for 1 month or more; more than 50% of property severely damaged)
☐ *Critical* (Injuries or illness resulting in permanent disability; shutdown of critical facilities for at least 2 weeks; 25% to 50% of property severely damaged)
☐ *Limited* (Temporary injuries; shutdown of critical facilities for 1-2 weeks; 10% to 25% of property severely damaged)
☐ *Negligible* (Injuries treatable with first aid; shutdown of critical facilities for 24 hours or less; less than 10% of property severely damaged)

Are any areas or facilities more likely to be affected (e.g., air, water, or land; infrastructure)? If so, which?

Fundamentals of Emergency Management

APPENDIX A: JOB AID

Job Aid 5.1: Hazard Analysis Worksheet (Continued)

Speed of Onset:

☐ *Minimal or no warning*
☐ *6 to 12 hours warning*
☐ *12 to 24 hours warning*
☐ *More than 24 hours warning*

Potential for Cascading Effects?

☐ No
☐ Yes. Specify effects:

Fundamentals of Emergency Management

Appendix B: Acronym List

Appendix B: Acronym List

CAR	Capabilities Assessment for Readiness
CERT	Community Emergency Response Team
DEOC	Department Emergency Operations Centers
DOD	Department of Defense
DOE	Department of Energy
DOT	Department of Transportation
DMAT	Disaster Medical Assistance Team
DMORT	Disaster Mortuary Team
EEI	Essential Elements of Information
EMA	Emergency Management Agency
EMAC	Emergency Mutual Aid Compacts
EMI	Emergency Management Institute
EMS	Emergency Medical Services
EOC	Emergency Operations Center
EOP	Emergency Operations Plan
EPA	Environmental Protection Agency
EPI	Emergency Public Information
ESFs	Emergency Support Functions
FEMA	Federal Emergency Management Agency
GAR	Governor's Authorized Representative
HHS	Department of Health and Human Services
HSPD	Homeland Security Presidential Directive
IACP	International Association of Chiefs of Police
IAP	Incident Action Plan
ICP	Incident Command Post
ICS	Incident Command System
JIC	Joint Information Center
NEST	Nuclear Emergency Support Team
NFA	National Fire Academy
NFPA	National Fire Protection Association
NIMS	National Incident Management System
NGO	Nongovernmental Organization
NRF	National Response Framework
NSDS	National Safety Data Sheet
NVOAD	National Voluntary Organizations Active in Disasters
PDS	Professional Development Series
PIO	Public Information Officer
SBA	Small Business Association
SOPs	Standard Operating Procedures

APPENDIX B: ACRONYM LIST

TTY — Text Telephone

USDA — U.S. Department of Agriculture

Appendix C: Emergency Supply Kit

APPENDIX C: EMERGENCY SUPPLY KIT

Emergency Supply Kit

Water

Store water in plastic containers such as soft drink bottles. Avoid using containers that will decompose or break, such as milk cartons or glass bottles. A normally active person needs to drink at least 2 quarts of water each day. Hot environments and intense physical activity can double that amount. Children, nursing mothers, and ill people will need more.

- Store 1 gallon of water per person per day (2 quarts for drinking, 2 quarts for food preparation/sanitation.)*
- Keep at least a 3-day supply of water for each person in your household.

If you have questions about the quality of the water, purify it before drinking. You can heat water to a rolling boil for 10 minutes or use commercial purification tablets to purify the water. You can also use household liquid chlorine bleach if it is pure, unscented 5.25% sodium hypochlorite. To purify water, use the following table as a guide:

Ratios for Purifying Water with Bleach

Water Quantity	Bleach Added
1 Quart	4 Drops
1 Gallon	16 Drops
5 Gallons	1 Teaspoon

After adding bleach, shake or stir the water container and let it stand 30 minutes before drinking.

Appendix C: Emergency Supply Kit

Emergency Supply Kit (Continued)

Food

Store at least a 3-day supply of nonperishable food. Select foods that require no refrigeration, preparation, or cooking and little or no water. If you must heat food, pack a can of Sterno®. Select food items that are compact and lightweight. *Include a selection of the following foods in your Emergency Supply Kit:

- Ready-to-eat canned meats, fruits, and vegetables
- Canned juices, milk, soup (if powdered, store extra water)
- Staples—sugar, salt, pepper
- High-energy foods—peanut butter, jelly, crackers, granola bars, trail mix
- Vitamins
- Foods for infants, elderly persons, or persons on special diets
- Comfort/stress foods—cookies, hard candy, sweetened cereals, lollipops, instant coffee, tea bags

First Aid Kit

Assemble a first aid kit for your home and one for each car. A first aid kit* should include:

- Sterile adhesive bandages in assorted sizes
- 2-inch sterile gauze pads (4-6)
- 4-inch sterile gauze pads (4-6)
- Hypoallergenic adhesive tape
- Triangular bandages (3)
- Needle
- Moistened towelettes
- Antiseptic
- Thermometer
- Tongue blades (2)
- Tube of petroleum jelly or other lubricant
- Assorted sizes of safety pins
- Cleaning agent/soap
- Latex gloves (2 pairs)
- Sunscreen
- 2-inch sterile roller bandages (3 rolls)
- 3-inch sterile roller bandages (3 rolls)
- Scissors
- Tweezers

Nonprescription Drugs

- Aspirin or nonaspirin pain reliever
- Antidiarrhea medication
- Antacid (for stomach upset)
- Syrup of Ipecac (used to induce vomiting if advised by the Poison Control Center)
- Laxative
- Activated charcoal (used if advised by the Poison Control Center)

Emergency Supply Kit (Continued)

Tools and Supplies

- Mess kits, or paper cups, plates and plastic utensils*
- Emergency preparedness manual*
- Battery-operated radio and extra batteries*
- Flashlight and extra batteries*
- Cash or traveler's checks, change*
- Nonelectric can opener, utility knife*
- Fire extinguisher: small canister, ABC type
- Tube tent
- Pliers
- Tape
- Compass
- Matches in a waterproof container
- Aluminum foil
- Plastic storage containers
- Signal flare
- Paper, pencil
- Needles, thread
- Medicine dropper
- Shutoff wrench, to turn off household gas and water
- Whistle
- Plastic sheeting
- Map of the area (for locating shelters)

Sanitation

- Toilet paper, towelettes*
- Soap, liquid detergent*
- Feminine supplies*
- Personal hygiene items*
- Plastic garbage bags, ties (for personal sanitation uses)
- Plastic bucket with tight lid
- Disinfectant
- Household chlorine bleach

Clothing and Bedding

*Include at least one complete change of clothing and footwear per person.

- Sturdy shoes or work boots*
- Rain gear*
- Blankets or sleeping bags*
- Hat and gloves
- Thermal underwear
- Sunglasses

Appendix C: Emergency Supply Kit

Emergency Supply Kit (Continued)

Special Items

Remember family members with special needs, such as infants and elderly or disabled persons.

For Baby*

- Formula
- Diapers
- Bottles
- Powdered milk
- Medications

For Adults*

- Heart and high blood pressure medication
- Insulin
- Prescription drugs
- Denture needs
- Contact lenses and supplies
- Extra eye glasses

- Entertainment—games and books
- Important Family Documents —keep these records in a waterproof, portable container
- Will, insurance policies, contracts, deeds, stocks and bonds
- Passports, social security cards, immunization records
- Bank account numbers
- Credit card account numbers and companies
- Inventory of valuable household goods, important telephone numbers
- Family records (birth, marriage, death certificates)

Appendix C: Emergency Supply Kit

For more information, go to http://www.ready.gov